DO NOT REMOVE
CARDS FROM POCKET

# BO RUDIN

# MAKING PAPER

*A Look into the History of an Ancient Craft*

RUDINS
Box 5058
S-162 05 Vällingby
Sweden

# CONTENTS

# PREFACE

We do not as a rule need much encouragement to talk about things that are of greater interest to us than most other things in this world. Yet this book would not have been published without some friendly advice and even more friendly pushing. "You really must do it!"

Well, if I must, I must.

The friendly pressure group consisted of fellow artists and papermakers, among them *Jordi Arkö* and *Richard Årlin* and last but not least the fantastic *Georg Anzelius*. Georg is now into his 74th year as a papermaker. He started in 1916 as an apprentice at Tumba Paper Mill near Stockholm and is still in full swing with lectures and demonstrations. The chapter "The Papermaker" deals with his lifelong dedication to papermaking. This rather lengthy interview took place in 1986 and is more or less an initiation for the rest of the book.

Since childhood I have nourished an interest in Japan and Japanese art and thanks to a grant from the *Foundation for Swedish Writers* it has been possible for me to visit the Kikuchi family of papermakers in Yamagata-machi. Among them I found a knowledge and a dedication to the art of papermaking equal to George's. My first contact in Japan, *Asao Shimura*, well known for his writings on the subject, also helped me a lot and encouraged the publishing of this book.

In addition to the above mentioned I would like to thank Mrs *Anna-Lisa Liljedahl* for giving me permission to use the late *Gösta Liljedahl's* text about "Watermarks", and *Arnulf Hongslo* for contributing "The Chemistry of Paper".

The following paper companies were kind enough to supply the paper for the book: Forenede Papir, Copenhagen, Denmark (Mischa 120 g).

Pappersgruppen, Stockholm (Macoprint 115 g for the colour sheet) and Tumba Paper Mill (Ingres 120 g for the jacket and Tre Kronor 120 g end papers).

May 1990

*Bo Rudin*

# PAPER OR NOT?

**Papyrus, tapa, amate, rice paper and parchment**

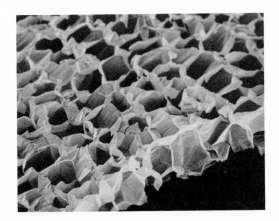

*Previous page: Rice paper as seen through an electronic microscope. The total width of the image equals approximately 0.01 mm. Photograph: Björn Hallström.*

Paper has been man's companion for more than 2,000 years. It is one of the simplest and yet one of the loveliest materials we know, and at the same time one of the least appreciated.

Paper is made by matting and intertwining vegetable fibres. Enormous technical advances have been made in the manner of its manufacture over the past few centuries, but fundamentally, the principle underlying the first paper we know of still prevails. The procedure falls into three states. First the raw material is turned into pulp. Then the fibres are dispersed in water and stirred into a slurry; this is transferred to a frame, through which the water is drained off, leaving the fibre layer behind. Lastly, the fibre layer is removed from the underlay, pressed, dried and turned into the finished product.

To qualify as paper, the raw material must consist of fibres which have been ground or otherwise crushed, so that each fibre constitutes a separate unit. This is what distinguishes genuine paper from *papyrus, tapa, amatl* and *"rice paper"*, for example.

Papyrus is the oldest writing material in existence. It was already being made by the Egyptians about 5,000 years ago, from the papyrus plant, *Cyperus papyrus*. This is a perennial grass species occurring in large parts of Africa and capable of growing up to six metres high. It is mostly found near rivers and on marshlands. This plant consists of an extensive rootstock system which produces buds in the form of long stalks. The stalks are triangular, narrowing towards the top and surmounted by a large, umbrella-shaped flowerhead. The papyrus stalk has no pulvini and it has just a few leaves, very low down near the root. The Egyptians quickly perceived the usefulness of this plant. Its beautiful flowerhead was used as an ornament, the root supplied both fuel and material for making things, while the stem was turned into rope and cable for boats or stout fabric for sails, sandals and matting. Dried stems were bundled together and used as rafts. Above all, though, papyrus was used as a writing material. Our word "paper" is derived from the Greek name of the plant, *papyros*.

The oldest of the papyrus rolls to have been discovered date back no less than about 5,000 years. Although the tools and methods then used were very primitive, it is an established fact that, in terms of quality, this writing material perfectly well bears comparison with the technically complicated products of our own age. Indeed, few present-day paper-makers can warrant their products for 5,000 years!

Until the 10th century A.D., papyrus was one of Egypt's biggest export articles, second only to linen textiles.

When, at the end of the 10th century, the Arabs introduced them to paper, it was not long before the Egyptians abandoned papyrus. Cultivation was discontinued, the papyrus plant eventually disappeared from Egypt and it was not until 1962 that cultivation was resumed by the Egyptian papyrus expert Hassan Ragab.

The oldest surviving description of papyrus production is by the Roman author and philosopher Pliny the Elder (A.D. 23-79). No Egyptian account has come down to us, presumably because the Egyptians were jealous of their trade secret: they had a monopoly of production.

Pliny writes that the stalks must be cut off near the root. The upper parts and the flowerhead, as well as the parts nearest the root, were discarded, being considered unfit for papyrus production. The remaining stalks were cut up into lengths of about half a metre and split down the middle. The parts nearest the middle, of course, were widest, and best for producing sheets with, while the narrowest parts on the outside of the stalk were discarded as being too hard and narrow. After the pieces had been cut into thin strips they were laid side by side on a board. They were then covered with a thin layer of slurry made from wheaten flour and the muddy water of the Nile with a little vinegar added.

Another layer of papyrus strips was superimposed on the first and the two layers were then pressed and hammered together. The result was a laminate which made an excellent writing material. Production-wise, however, it was completely different from paper.

When Dr Hassan Ragab, whom we have already mentioned, began studying the history of papyrus in 1962, his first problem was to find seedlings for cultivation in Egypt, because the plant was extinct there. To find what he was looking for he had to go as far afield as Sudan, Egypt's southern neighbour. Dr Ragab now runs an institute of papyrus research in Cairo, and the institute's papyrus plantations are the largest in the world.

Hassan Ragab does not have much time for Pliny's description. He discovered that, translated into practice, the methods described in Pliny's book are soon found to be unworkable. He puts this down to Pliny never having experimented on his own; everything he wrote was hearsay, gathered from informants who balked at revealing too many secrets of their lucrative craft. Ragab's own methods have been evolved through many years' experimentation at his institute.

When harvesting papyrus, the first thing to do is to cut off the flowerhead, so that it will not tear down any neighbouring plants when the stem is felled. On no account must the stem be damaged. If it is bent or otherwise harmed, brown

*Left: Cutting the papyrus into strips. Right: Wetting and rolling the dried papyrus.
Photograph: Hassan Ragab.*

stains will appear in the otherwise chalk-white core, and the finished sheet will
be flawed.

Little more than half a metre of the stem is used for production purposes, viz
the part nearest the ground. The peel or bark is removed with a sharp knife,
exposing the core. This is then cut lengthwise into thin parallel strips.

During the harvest season, between June and September, fresh papyrus strips
can be used immediately for sheet production, but they can also be dried and
stored for subsequent use.

The procedure for making sheets is as follows:

A piece of cotton fabric is laid over a dry felt on the workbench. The first layer
of the sheet is begun by putting up a 40 cm long strip of papyrus horizontally. The
next strip is positioned parallel to the first, with an overlap of a few mm. The
remaining horizontal strips are similarly positioned until the sheet has been built
up to the requisite height, which is about 30 cm.

The strips for the next layer, which are 30 cm long, are laid at right angles to
the first, with the same overlap, until the full width - 40 cm - has been attained.
The actual wet sheet has now been fashioned, and it is now covered with a new
piece of cotton fabric, so as to keep the strips in position while work proceeds.
Another felt is laid over the cotton to absorb the moisture when the sheet is being
pressed.

After a bundle or post of about 10-15 sheets, with felt on both sides of each
sheet, has been put in position, the moisture is squeezed out of the wet strips and

absorbed by the dry felts. Ordinary screw presses are used, and the post is left in them for two hours during the first pressing operation. After this the press is opened and the felts replaced, without displacing the cotton fabric. After another ten hours the procedure is repeated and everything is left in the press for a further 24 hours. The papyrus sheets are then completely dry and their thickness about 5 per cent that of the original strip. Since the surface area of the papyrus sheet does not differ appreciably from that of the sheet of newly laid strips, one might wonder where the material inside what began as 4 mm thick strips has got to. It is equally natural to ask what holds the sheet together and keeps it from disintegrating again after drying.

Ragab's explanation is that the papyrus core contains saccharids which glue the vegetable fragments together during pressing, with the result that all the air ducts of the hollow core form a kind of "dovetail" during pressing, and in this way hold the sheet together. Pliny's theory of the muddy water of the Nile serving as a glue was probably just a tall story from his unwilling informants.

*Tapa* (the name comes from the Polynesian word for bark-paper) is a fabric-like paper made from the inner bark of the paper mulberry tree. Other tree species are also used, e.g. the fig and breadfruit trees. When these trees have grown large enough, about 5 metres high and 20 cm in diameter, they are cut at the base and divided up into convenient half-metre pieces. The pieces are then bundled up and put in running water until the outer bark feels soft. It is then scraped away with tools made from ground shells, exposing the white inner bark.

The next stage is to make an incision along the length of the wood and to remove the inner bark in thin strips, which are then put out to dry in the sunshine. After drying the strips are coiled with the concave side outwards. This causes them to straighten out, which facilitates the ensuing process.

Before work begins on beating the strips with wooden clubs, they are soaked in water. During the beating operation the strips expand to about five times their original width. Any holes are mended with pieces of bark beaten into the strips.

The thin, wide strips are then re-dried and the finished material coloured with vegetable dyes and sewn together into garments or used for various magical rites.

There is no firm evidence concerning the origins of tapa, but certain references suggest that this fabric was being used in South-East Asia as long ago as about 4000 B.C. Fragments discovered during excavations in Peru suggest that tapa was being used there in about 2100 B.C. It is believed to have been introduced in the Pacific region at about the beginning of the Christian era. Unfortunately this material is sensitive to the damp of the tropics, with the result that very few specimens indeed of earlier tapa work have survived.

Tapa is widely used as writing and art material, for clothing, bedding and also for gifts and ceremonially.

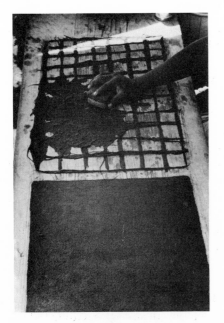 

*Left: Amate strips placed crosswise and beaten. Right: Finished strips drying in the sun, with raw fibre in the foreground. Photographs: Ulrika Hembjer*

*Huun* and *Amate* were the writing materials of the Maya and Aztecs. Their technique of sheet forming reminds one of papyrus production, but the processing of the raw material is completely different, having more in common with tapa.

Huun was a beaten bark material used by the Maya as writing material and for artistic purposes. Its development matched the intellectual progress of the Maya and their hieroglyphic script. During the last great period of Maya civilisation, c.900, the Maya had begun folding sheets of huun together into books. The *Codex Dresdensis* is an almanac consisting of 45 pages, made sometime between 900 and 1100. Spanish chronicles tell us that the Codex Dresdensis was by no means the only book to have been produced by the Maya people; they had many books, in fact complete libraries. A Spanish missionary, Diego de Landa, from the monastery of Izamal in Yucatan, wrote in the 16th century: "We found a great number of books written with their characters, and because they contain nothing but superstitions and falsehoods about the devil, we burned them all......"

But it was the Aztecs, succeeding the Maya, who developed the art of converting bark into writing material. They called their material amatl. According to an eye witness description by Dr Francisco Hernandez, a Spanish scientist who went on an expedition to Mexico in 1570, the production process was as follows. "They chop the largest branches off the tree (*amaquahuitl,* literally 'paper tree')...... and leave them to soak overnight. The next day the outer bark is pulled off and the inner bark is cut away in long strips. These are then put out on boards of hardwood and beaten together into large sheets which, after they have dried, are polished with a smooth stone. The sheets resemble our own paper, but are somewhat whiter and thicker."

The Otomi Indians of southern Mexico still make amate in a similar fashion. The only difference is that they boil the bark before beating it. After boiling the strips are put on a board at intervals of a few cm, and after beating, when the strips have expanded sideways, the result is a large sheet which is then put out to dry in the sun, still on its board.

It does not take very much imagination to perceive the similarity between the methods employed for producing papyrus, tapa, amate and huun. None of these products meets the criteria of "genuine" paper, but they need not be inferior for that.

Unlike papyrus, amate is used mostly for magical rites. Silhouettes of white amate are cut out and fastened to a brown underlay, or vice versa. Otomi Indians today also use their amate as a ground for brilliantly coloured paintings.

*Rice paper* has nothing to do with real paper or with any of the substances mentioned above. It is a paper-like material made from the spongy pith of branches and trunks of the rice-paper plant, *Tetrapanax papyriferus.* Nobody knows when rice paper was first manufactured, but production of paper from pith is mentioned by a Chinese author writing in 1634.

Rice paper has been put to only very limited use as a writing material. It has mainly been used as watercolour paper and in the production of artificial flowers. As an export article it was sold both in the USA and Britain and everywhere in the Orient during the 19th and early 20th centuries. It is still made in Taiwan, though on a limited scale. Rice paper is white and soft, with a velvety surface, and it is made in small sheets only.

The name rice paper has been a cause of almost utter confusion. In Britain it acquired this name because the British believed it was made from rice stalks or even from rice itself. Later on a thin, white paper made from the bast of the paper mulberry tree in Japan came to be known as rice paper. Finally, the Chinese produce a thin white paper of bleached rice stalks which is also called rice paper.

But the only plant used for manufacturing rice paper is the above mentioned rice paper plant which is a shrub or small tree. The part actually used is the pith of thick branches or the trunk, taken from either wild or cultivated plants. The pith has to be soft, but not as "grainy" as papyrus, otherwise it is impossible to produce the thin, veneer-like sheets. The procedure is as follows:

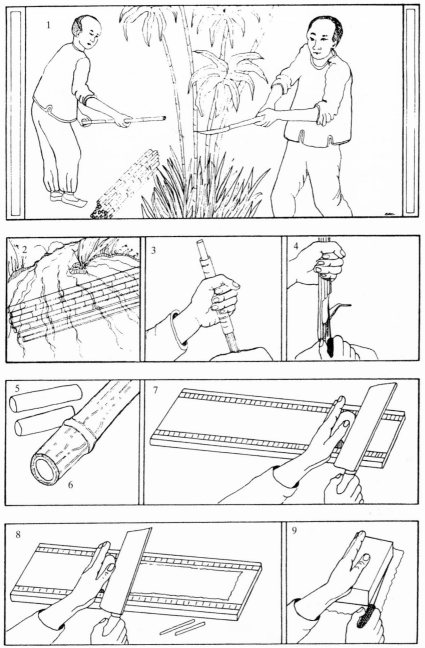

*Rice paper production. 1. Harvesting. 2. Soaking the stems. 3. Extracting the pith, method 1. 4. Extracting the pith, method 2. 5. Natural drying. 6. Drying in a bamboo tube. 7, 8. Sheet formation. 9. Trimming the edges. Illustrations: Lilian Bell (1)*

The plant is best harvested in wintertime, but, whether wild or cultivated, it can be cut at any time of the year. In Taiwan, where the plant is cultivated, the first branches are harvested after three years' growth. The second harvest comes after the fifth year and the last one after seven years. The final harvest includes the whole tree, after which new shoots are planted. Cultivated trees yield the best-quality pith.

The branches or trunks are then cut into lengths of between 2.5 and 3 metres and put to soak in running water or in a bath of fresh water, where they are left for a few days until the pith has disintegrated somewhat, which makes it easier to remove. The water in the bath is replaced once a day, as a precaution against mould.

There are several methods for extracting the pith. One of them is to push it out with a wooden plug. This is rather an awkward business if the branch is crooked. Eventually, though, the pith emerges at almost the speed of a champagne cork.

Another method is based on carefully carving away the outer bark and the woody parts of the plant. This is not as hard as it sounds; a potato peeler can be used for paring away the wood next to the pith. The method is not recommended, though, because this way it is hard to achieve pith of the ideal cylindrical shape.

The pith has to be dried before it can be put to any further use. It is put out to dry immediately after it has left the branch, and after several days in the sun it is put away dry.

The length of dry pith, about 25 mm in diameter, are cut into 7-10 cm pieces. When they are to be cut into sheets, they are laid over a chopping board with a brass strip on each side. An incision is then made over the pith with a sharp knife. This knife is then placed across the chopping board, so that it rests on the brass strips. The latter raise the plate about 0.5 mm above the board. Starting with the incision made earlier, the pith is now rotated against the knife, which then cuts off a long strip.

No further drying is needed, and so the strips or rolls of pith are stored until enough of them have been collected to be cut into small sheets.

*Parchment* is a classical writing material which, again is unrelated to paper. Tradition tells us that the name comes from the ancient city of Pergamon. When King Ptolemy of Egypt prohibited exports of papyrus, so as to frustrate the growth of the Pergamon library, the people of Pergamon began using parchment as a book material. Ever since about 100 B.C. parchment has been used instead of papyrus, which is less durable and a lot more expensive.

The raw material for parchment consisted of sheep skins, goat skins or calf skins which were dried on a frame, scraped and polished with a pumice-stone or chalk and, finally, trimmed at the edges. The pieces of parchment were lined up

*Parchment maker. From Hartmann Schopper: Panoplia Omnium ... Artium, 1568.*

horizontally and vertically (for margins), folded in the middle and assembled into one or more gatherings, containing up to twelve sheets. This was the origin of the codex, the commonest type of book after the 4th century. Carolingian parchment is particularly famous for its beauty.

9

Modern parchment is made from the skins of sheep, calf, goat and lamb. This last mentioned is considered extra fine and is known as vellum. After the skins have been cleansed from hair, they are put into a clear solution of lime for two or three weeks, stretched on frames and air-dried. After this they are rubbed with pumice-stone in powdered chalk. For a glossy surface, parchment is greased with egg white and polished.

Nowadays parchment is mostly used as book-binding material for fine bindings and for illuminated addresses and suchlike.

# PAPER'S ODYSSEY

**The thousand-year progress from China to Europe**

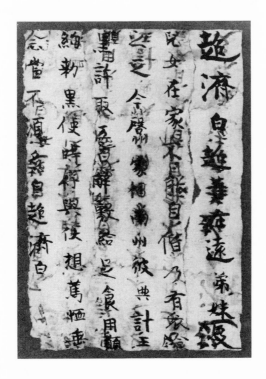

Paper was invented, before the Christian era, in China, a country where learning and the written word were valued highly. About 2,000 years ago a need developed there for a new writing material to replace the sticks of wood and bamboo employed previously. With the widespread introduction of a new writing implement - the brush - people began to write on silk. Compared with wood and bamboo, silk had the advantage of being light and pliable and an admirable vehicle for calligraphic script. But it had one disadvantage: it was costly.

Finding a substitute with properties resembling those of silk cannot have been easy, but it is clear that experiments with new materials began a few centuries before Christ. Nobody knows exactly when the first paper appeared in China, but during excavations round the dry lake of Lop-Nor in 1901, the explorer Sven Hedin found fragments of paper which can be dated to about 250 A.D. This paper, long believed to be among the oldest in the world, is now in the Ethnographic Museum of Sweden, Stockholm. Even older paper, believed to date from about 140-87 B.C., was found in Xian in 1957. It was first believed that this paper had been made from silk fibre, and it was not until 1964 that microscopic studies identified the original material as hemp fibre.

This discovery of paper antedating the Christian era calls for a slight revision of the history of paper as received before 1964. Tradition has it that the noble art of making paper was invented by a Chinese eunuch by the name of Ts'ai Lun. For he is the first paper-maker to be mentioned in Chinese history books. In 105 A.D. he reported to the Emperor Ho-ti that he had successfully produced a writing material from hemp, bast, discarded fishing nets and cloth. The Emperor greatly appreciated this ingenuity, and the new material was duly adopted. All we know concerning Ts'ai Lun's subsequent career is that he became embroiled in a court intrigue and, to avoid being put on trial, went home, took a bath, combed his hair and drank poison.

As regards the method of producing the pulp, we know that the raw material was beaten in stone mortars together with water. Mortars of this kind are mentioned in the story of Ts'ai Lun and the American paper expert Dard Hunter found similar mortars on his travels in China about 50 years ago. The same Dard Hunter is also convinced that the earliest sheets of paper were made by diluting the pulp with water until one obtained a fibrous slurry, which was then poured

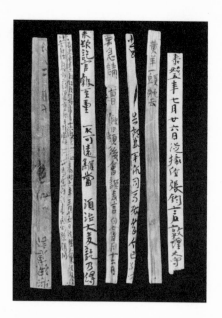

*Left: Prior to the invention of paper, the Chinese wrote on bamboo sticks. Photograph: The Ethnographic Museum of Sweden. Right: Chinese accordion or leporello binding.*

over a screen set up in a frame. The Chinese were expert felt-makers, a craft which may have enhanced their understanding of the matting of fibres.

This technique is the oldest known method of paper-making, and the principle - macerating a material into fibre and then producing a new material, paper - has never changed. The old method, however, has one drawback. The sheet cannot be removed from the frame before it dries. Sometime in the early history of paper, however, somebody hit on the idea of constructing a frame in which the stretched fabric was replaced with a mat-like structure. The mat was made of thin bamboo laths and merely rested on a wooden frame. By holding the mat onto the frame with two sticks, one could dip it into the fibrous slurry and remove it with a layer of pulp on top. The flexibility of the mat - it could be rolled up - made it possible to transfer the fibrous layer to a new underlay for drying. Work could then begin on a new sheet, and so on. The earlier simple frame had now become a paper mould, and a new method was developed which, eventually, was to conquer the world.

Paper was used above all for the propagation of ideas. All early finds bear witness to paper as a vehicle of written information. Because thin paper could only be used on one side, bamboo writing sticks were still being used as late as the fifth

*Chinese paper-making. From T'ien-kung k'aiwu, a dissertation on crafts by Sung Ying-
hsing, 1634. Top left: Steaming bast. Top right: Paper formation. Bottom left: Pressing.
Bottom right: Drying against a heated wall.*

century. By the end of that century, however, they had been wholly superseded by paper.

Paper really came into its own with the invention of printing. When, in the 7th century A.D., the holy scriptures carved on tablets of stone were to be made available to the people, the method was adopted of colouring the tablets and making impressions from them on paper. Chinese paper, being strong and pliable, was eminently suitable for this method of reproduction, and it was then only a short step to printing engraved pictures or characters from woodblocks. Thus the Chinese art of printing came into being. Paper and the nature of Chinese script decided the form. The Chinese, with their large number of written characters, roughly varying between 4,000 and 6,000 in a normal book, do not have the same need of movable type as the Western peoples, with their no more than 30 constantly recurring characters.

The oldest books were made in scroll form, and were printed on one side of the paper only. The leporello (accordion) book, in which the sheets are stuck together into a long strip and folded instead of rolled, is an intermediate form between the roll and the paged book. This device made books easier for the reader to handle.

Because of printing, China soon became a land of books. Art and literature flourished during the T'ang period, when close studies of classical literature and various branches of science were a necessary qualification for public service. Paper-making in China was kept a closely guarded secret from foreigners until well into the 6th century, when the craft reached Korea, proceeding from there to Japan at the beginning of the 7th century. It was Buddhist monks who took with them manuscripts on paper made from the bark of the paper-mulberry tree. Ancient annals record that the Japanese began making paper in 610 A.D., and they are believed to have been taught the art by a Buddhist monk called Doncho. He was a man of many parts who, in addition to making paper, was also able to make ink and paint. In addition, he was skilled in medicine. Eventually this monk became physician in ordinary to Prince Shotoku and an influential counsellor. Paper may have originated in China, but the first block printing took place in Japan. The oldest known example of this printing method comes from the reign of Prince Shotoku. In 770 he had strips of Buddhist sutras printed and "one million" copies put into wooden urns resembling pagodas.

Once the Chinese paper-makers had shared their knowledge with the Japanese, the art rapidly spread throughout the island kingdom. During the Nara Period (708-806 A.D.) paper was made in nine provinces, and in the Heian Period (806-1155 A.D.) there were 40 provinces practising the craft. Some time in the 9th century Zushoryo (the department of the library where the books, drawings and paintings of the Emperors were kept) had begun making paper under the direction of four highly trained craftsmen. This enterprise was begun to influence papermaking in Japan, and in 807 the art of papermaking was introduced in the neighbouring city of Kyoto, regarded as the artistic centre of Japan.

During the Heian Period, the central administration declined and the Zushoryo ceased to exert any influence. The slave-like guild of paper-makers, which had hitherto been segregated, mingled with ordinary people and, before long, the imperial papermaking staff was reduced in both number and ability.

Many of the paper-makers, however, were taken on by private landlords who had started papermills of their own.

Up to this time, paper in Japan was made almost entirely from three raw materials: *gampi* (Wikstroemia canescens) and *hemp* (Cannabis sativa) together with *kozo* (Broussonetia papyrifera). As early as 1031, however, there are references to waste paper being used in the production of new sheets. The Chinese had used this method before, and since the Japanese had learned so much from them, it is reasonable to suppose that the re-use of waste paper was no exception. The re-cycled material was charged with ink and pigment, with the result that the new paper acquired a greyish hue, but it was very much in demand for all that.

It is difficult to arrive at an exact date when *mitsumata* (Edgeworthia papyrifera) began to be used, but there are references in 1597 to a family having been granted permission to collect mitsumata bark.

Paper and the craftsmen capable of producing these thin, beautiful sheets were greatly venerated by the Chinese, Japanese and Koreans. Probably, though, there is no country in the world which can rival the ingenuity of the Japanese in finding uses for paper: as writing material, for calligraphy, as fans, garments, dolls and as an important component of Japanese houses - the translucent *shoji* which serves as a window, and the opaque *fusuma,* which makes a partition wall.

Of necessity, most paper used today is machine-made, and the same goes for Japan. Machinery, moreover, can produce imitations of hand-made paper, thus reducing the market for the latter. The great industrialisation of Japanese paper-making began after the Russo-Japanese War in 1905, and the hand paper-makers found it increasingly difficult to hold their own against the machines. They tried introducing some mechanical operations in order to compete, but mechanisation was not really compatible with the traditional production of beautiful sheets and the use of the old raw materials mitsumata, kozo and gampi.

During the present century, production of hand-made paper has fallen off dramatically in Japan. In 1901 there were 68,562 households producing paper. By 1941 this number had fallen to 13,577, in 1973 it was 886 and in 1979 it was 697. The decline has decelerated in recent years and conditions have levelled out

19

somewhat, which can justify hopes for the future of the craft.

In the closing years of the 6th century there began in China a period of great achievement in all fields, and the country had trading links with both Arab and Jewish merchants. Caravans moved through the mountain passes between China and Western Asia along the re-opened Silk Road. Chinese ships sailed to the Persian Gulf, and there was an Arab trading office in Canton during the mid-7th century. By then the Arabs were already importing paper from China, and paper money was probably being used as well. But it was not through peaceful trading that the Arabs were to learn the art of paper-making. During the 7th century the Chinese Empire had expanded to the region of Pamir. Meantime, in the west, a new world power had come into being: Islam, which by the early years of the 18th century had expanded as far as the outposts of the Chinese Empire in Eastern Turkestan. In a war between the Chinese and a Turkestan vassal, the latter was helped by the Arab governor of Samarkand. In July 751 a great battle was fought by the River Thalas. The Chinese were defeated and many of their troops captured. The captives included a number of paper-makers who, in an attempt to secure their release, began practising their craft. And so the art of paper-making came to the Arab world.

From the eastern fringes of the empire, the art spread slowly to other regions. Until now, papyrus had been the writing material used in the Arab world, but papyrus sales had begun to decline, especially under the impact of parchment, the dominant material in the Roman literary world from the 4th century A.D. Paper began to supersede papyrus, production of papyrus declined more and more, and by the mid-10th century it had virtually lapsed. Paper was victorious.

Production of paper in Samarkand was facilitated by abundant supplies of flax and hemp and by the existence of innumerable irrigation channels offering a supply of pure water. Samarkand paper eventually came to rank equal with the Chinese product, and for a long time it remained a Samarkand monopoly.

Then, during the 10th century, paper advanced to the eastern Mediterranean countries: Syria, Palestine and Egypt. During the 11th century, paper-making gained a firm foothold in Egypt, which also ha a good supply of the finest raw material - linen.

It took about 400 years for the art of paper-making to traverse the Arab world, from Samarkand to Spain and Morocco. In Spain, conquered by the Arabs in the early 8th century, paper was known by about the end of the 9th century. Spain also has the oldest surviving paper in Europe: a document dated 1009. An Arab travelogue records that in Xativa, south of Valencia, a paper was made the like of which was not to be seen anywhere else in the civilised world. Paper-making continued here under Moorish rule until 1244.

The production of paper had a galvanising effect on intellectual activity within the Moslem world. It led, not least, to the development of an admirable calligraphy, by no means inferior to that of the Chinese and the Japanese. The profession of the writer was a respected one, and schools were started where salaried scholars gave lectures to hundreds of students. These schools also had librarires, and at a time when a monastic library in Europe ranked as a large one if it had 100 volumes, the library of the Khalif Al Aziz in Cairo contained an estimated 150,000. Various sciences flourished in the Arab empire. An abundant literature came into being, and the works of the classical authors were distributed in large editions. This rich diversity of cultural development could not have been achieved without paper.

The Arabs, however, did not advance the art of paper-making in a technical sense. They kept to the manufacturing methods which the Chinese had taught them. The raw material (hemp and rope or linen rags) was cut up with sickles, washed in pure water and macerated in stone mortars with water added. The moulds were of the same kind as those used by the Chinese, i.e. mats of plaited rushes. The paper-maker smoothed the slurry over the mould with his hand. The fibrous layer was then couched on a board and then smeared up on a wall to dry. Afterwards the dry sheet was sized with rice starch and then dyed and glazed.

Arab paper-making, like its East-Asian counterpart, was mostly a domestic industry. The quality of Arab paper deteriorated during the 15th century and output declined. Paper now had to be obtained elsewhere, and "Frankish" paper was imported, mostly from Italy. The main European languages, however, have incorporated a memento of the heyday of Arab paper-making. A bundle of paper is called *ris* in Swedish (from the Arabic *risma*, meaning bundle or packet). In Spanish this is *rezma*, in Italian *risma*, in French *rame,* in German *Ries*, in Dutch *riem* and in English *ream.*

Paper began to be used in Christian Europe during the mid-11th century. It arrived by three main routes, via Spain, Sicily and Constantinople, through the agency of the Arabs. They had taken over their Chinese mentors' name for paper, *kog-dz* (paper made from the bark of the mulberry tree). In Arabic this became *kâghid.* The Europeans, however, employed the Graeco-Egyptian name for writing material, *papyros,* and it was not until production of papyrus had ceased that *papyros* came to be applied solely to paper.

Parchment, however, still occupied a strong position, especially where official document were concerned, and its survival was assured for the time being by the fluctuating quality of paper.

Paper-making in Spain was mostly in the hands of the Jews. When the Almohades gained ascendancy in the Moslem parts of Spain, the Jews were subjected to persecution and fled to the Christian kingdoms of northern Spain. Paper was already being succssfully made there - for the first time in Christian

*This page. A. The paper-maker (vatman) stands at his vat, lifting his mould with a layer of
pulp, ready to pass to the coucher. B. The vat, where the pulp is diluted with water and
heated, b aperture of the hearth which heats the vat. C. Deckle. D. D mould viewed from
two angles. E. Nut at the top of the press. F. The coucher receives the mould with the wet
sheet and couches it on the wet felt. f Wooden press-board. G. A post of wet sheets,
separated by wet felts. g Hooked stick for pulling the bundle of paper to press. H. H Press
for squeezing the water out of the bundle of paper. 1, 2, 3 Wooden shimmies for equalising
the pressure in the press. 5 Board on which the mould is pushed. 6 Drain which the vat man
puts his mould on. 7, 8 Supports for the mould I The lay boy, who removes the felts from
the sheets and gives them back to the coucher. K Lay boy, putting the newly-pressed sheets
into the lay stool. L Dry press. M Copper pan for puring new pulp into the vat. N Stick to
keep the press locked. O Hearth. From de la Lande: L'Art de Faire le Papier, 1698.*

*Next page, top. Sorting and fermenting rag after la Lande 1698. A, A, A are large chests or
boxes divided into three sections for different grades of rag. 1, 2, 3 boxes for fine, semi-fine
and coarse rags, B, B, B women sorting rags. 1, 2 knives used by women for cutting up rags.
C separate picture of one of these knives. 4, 4, 4 stationary rag cutters. D pile of excessively
coarse rags discarded by the women. E hole through which the rags are thrown down into
a special room for fermentation. LM, MN, NO width of the rag boxes. Lower picture: A,
A gutter supplying the fermentation tank with water. B wooden vat for collecting water for
the fermentation tank. C stone tank for fermentation. D stone chest where the rags are cut
up. E cutting instrument installed in the same. G bucket for carrying rags. H piles of rags
fermenting in one corner of the room. 1, 2, 3 rags falling from the upper department to the
lower.*

Europe - during the 12th century. As the Christian sphere in Spain expanded, a growing proportion of Moorish paper production fell into Christian hands. Spanish paper in the 13th century was of high quality and was exported to Sicily, Italy, France and England. Quality declined during the 14th century and the religious persecutions of the 15th forced innumerable Jewish paper-makers to flee to France. Spanish paper-making receded heavily at this time.

The Italian towns were already conducting a considerable amount of trade with the countries round about the Mediterranean during the early medieval period, and they learned the art of paper-making through their numerous trading offices abroad. Paper documents are extant from the 12th century, but the Italians themselves did not begin manufacturing paper until the second half of the 13th century. The first reliably recorded papermill was started - and is still in operation - in Fabriano, in the province of Ancona on the eastern slopes of the Apennines.

The Italians brought manufacturing techniques to a very high pitch and exported a great deal of their output. The principal paper-making cities were Venice, Milan and Genoa, and the fact is that for a long time Italian paper had a monopoly of the European, Levantine and African markets.

Paper began to be used in France in the early years of the 13th century. During the first half of the 13th century, most of it was of Spanish origin, but Italian paper become increasingly predominant during the second half. We do not know of any French papermill before 1338, when one existed in Troyes. Presumably the French learned the art of paper-making from the Italians, but there is always the possibility of the French crusaders having picked up the art in the Orient and brought it back home with them. Just as in Italy, it was above all the ecclesiastical authorities and the monasteries which took the lead in paper production. By about 1470 France was not only self-sufficient in paper but was competing successfully with Italy on the German market.

Germany and Switzerland did not begin using paper until about 100 years after France. In the mid-14th century diplomas in Germany began to be written on paper, which thus gained acceptance for official purposes. The Germans, however, found it an excessively perishable material, and it was not until 1549 that the Council of Augsburg sanctioned the use of paper for documents previously written on parchment.

Germany began making paper in about 1390, Switzerland about 20 years later, in 1411. The first German papermill was founded by Ulman Stromer, a Nuremberg councillor and merchant. Assisted by Italian mill-wrights, he converted a flour mill outside the city into a papermill. Just as in other parts of Europe, the paper-makers were forced to take an oath that they would not divulge the secret of the art, but this was not a very effective means of preventing

24

the establishment of papermills, now that Germany was surrounded by other paper-making countries. The art of paper-making was rapidly improved in Germany, and its greatest period of achievement came during the 16th and 17th centuries. Experts maintain that the paper used by Gutenberg for his 42-line Bible has never been surpassed in Europe.

Last of all, paper came to Austria, Bohemia and Hungary, from either Germany or Italy. It was already being used in the chancery of the King of Hungary by the end of the 13th century, and there are Austrian and Bohemian paper documents extant from the early years of the fourteenth. Quite a long time was to pass before these countries began making their own paper: Austria in 1469, Bohemia in 1499 and Hungary not until 1546.

Holland had been using paper since the beginning of the 14th century, but the first Dutch papermill was only founded in 1586, in Dordrecht. During the 17th century, Holland became the leading paper country in Europe, thanks partly to the skilled paper-makers who took refuge there from Louis XIV'S persecution of the Huguenots. Dutch paper became synonymous with quality, and Holland exported any amount of paper, mostly to the countries of northern Europe.

The Dutch also acquired or otherwise laid hands on any number of papermills outside their own frontiers, especially in France. They maintained their leading position in European paper-making for as long as the industry remained a craft. Rembrandt, for example, printed his works on van Gelder or Japanese paper.

Paper came to England during the first half of the 13th century, and from the end of that century onwards it was imported from France. It was not until 1588 that the English began making their own paper, although a papermill had been founded at Dartford during the closing years of the 15th century by a German called John Spilman. In fact, England never became self-sufficient in paper and imported large quantities, mostly from Holland.

The arrival of paper in the Scandinavian countries was comparatively late, but its acceptance was all the more rapid. Paper documents were first mentioned in the mid-14th century, and the oldest surviving specimens are from 1345 in Sweden, 1364 in Norway and 1367 in Denmark. By the 15th century, paper was being widely used for documents and writings of different kinds, but none was manufactured in Sweden for another century or so. This may seem curious, in view of the highly favourable conditions for paper-making in Sweden. Both water power and pure water were abundantly available, but throughout the Middle Ages both royal chanceries and cathedral chapters bought their paper - expensively - from other countries.

Paper probably reached the eastern parts of Europe by way of Constantinople, but by the end of the Middle Ages this trade was controlled by the Germans. Krakow had a papermill in 1491, there was one near Lemberg in 1533 and another near Luck in 1570. An unreliable source has it that the first Russian papermill was established in 1576, but paper-making is not firmly recorded until 1650.

Finally, the art of paper-making crossed over to America. It was Spanish paper-makers who, in 1680, founded the New World's first papermill, near the capital of Mexico. Another century was to elapse before paper began to be made in north America. In 1690 a German immigrant with the anglicised name of William Rittenhouse founded a papermill at Germantown, near Philadelphia, and this is where the first American paper-makers were trained.

It was not until 1856, however, that the art of paper-making crossed the continent of America and arrived in California. Canada began producing paper in 1803, Australia in 1868, near Melbourne.

Returing to Sweden, the first Swedish papermill is said to have been established to meet a sudden upsurge of demand - for printing. This mill was founded, near Linköping, by the dynamic Bishop Brask. During the 1560s there are references to one Torbjörn Klockare ("Torbjörn the Sexton") having founded, by command of King Gustav Vasa, a papermill at Norrström, Stockholm, which kept going for a decade or so. Paper-making was apparently established in Denmark at about the same time. Tycho Brahe, who himself founded a papermill on the island of Ven in Sweden, tells us that the credit for this achievement must go to his uncle, Sten Bille.

By then, according to a letter which Tycho Brahe wrote in 1589 to Christopher Rotmannius, his own papermill was in full swing. Paper-making is referred to again in 1591, and a year or so later we are told that Brahe has been granted a permit to purchase linen rags. Part of the letter reads as follows:

"The mill I mentioned has already been completed. A pond, enclosed by a high, broad bank, maintains a constant flow of water during both the warm and cold seasons of the year. The wheel, 12 ells in diameter, is driven by the smallest possible quantity of water and provides a power source for two other factories apart from the papermill."

Bille founded his papermill at Blekeskanamölla under Knutstorp in Skåne (Scania), now Sweden's southern-most province but at that time part of Denmark. That mills is presumed to be the precursor of the Herrevadskloster mill, which uses various forms of an HWC ligature as its watermark. Herrevadskloster in turn is the precursor of the later so famous Scanian Klippan papermill, although their locations were probably not identical.

The then County of Herrevadskloster occupied the central reach of the Rönneån River in Skåne. Water power had long been used not only for grinding corn but also for industrial purposes, and the fast-flowing Rönneån River made an ideal power source. The county with its great industrial potential was held in

fee by Sten Bille, a scholar and a benefactor of the sciences who was also
interested in a variety of industrial ventures. He helped to found several lime-
works and glassworks, and he also founded a papermill.

Our knowledge of that papermill is confined to references made by Tycho
Brahe and King Fredrik II of Denmark, and also to the few specimens of his
output now extant in the archives.

In 1573 Tycho Brahe celebrated his uncle Sten Bille, in a verse published in
Latin. This was the year after Brahe had discovered the new star which was to
immortalise his name in astronomy. The last two lines read:

The renowned art of paper-making, hitherto practised by none in our country,
was brought to life by Bille.

Three years later, Bille's papermill cropped up again. Skåne, as we have
already seen, was at this time a Danish province, and King Fredrik II wrote to
Bille, asking him to send his paper-maker across for a few weeks until another
could be called in from Germany, the reasong being that the King was planning
to start a papermill on the island of Zealand, near Copenhagen. The master
paper-maker arrived in April, and that autumn "Christoffer Rotther
Papirmager" received an old daler to pay for a silver crowned F and S (Fridericus
Secundus), "to be printed on paper when it is made".

A watermark, in other words. Sten Bille's own watermark displays a circle
divided by a cross, the stem of which reaches above the circle, with the letters SB
to either side.

Sixty-odd years after Bille's papermill, we heard of another papermill in
Skåne, once again in Herrevadskloster County. This is a continuation of the
earlier one, and so in this way the Klippan Mill is a descendant of Sten Bille's.

On 15th September 1651, Elias Biener, a clerk with the East India Company,
and a merchant with the name of Mattias Smidt were granted permission to start
a papermill at Stackarp. The two co-owners were awarded generous privileges by
King Christian IV. They were to use the mill without paying any charges to the
Crown for 15 years after its completion, and subsequently only a "cheap ground
rent" would be demanded of them.

Elias Biener proved unable to procure the necessary capital, and so Smidt had
to take charge of the enterprise from the very beginning. He, however, was no
expert on paper-making, and he preferred to lease the mill to a master paper-
maker on payment of an annual charge, a "mill rent" as it was called. After
signing a contract with Jochum Lemchen, a compatriot, Smidt returned to his
native Germany.

The mill was expensive to build, and Mattias Smidt had borrowed money on
the security of it. He frequently fell into arrears, and litigation followed. Mattias
Smidt died while visiting Denmark to look into things, and the following year the

 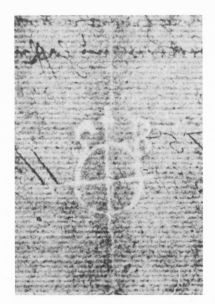

*Two old Swedish watermarks. Left: Herrevadskloster (HVC). Right: That of Sten Bille, who founded Sweden's first papermill (SB).*

mill was operated by a master paper-maker acting on behalf of his creditors. The latter fell out among themselves and in 1651, after further law suits, the papermill was declared forfeit to the Crown. Two months after this judgement, the Lord Lieutenant of Herrevadskloster assigned the mill on a lease to Jesper Krefting, a Lübeck merchant, who promised "within three months to put the mill in order and make paper good enough to be used by the King's Chancery and Treasury".

The Charter also confirms the emblem of the mill, a ligature of the letters HVC underneath a crown, HVC being short for HerreVadsCloster. Before long the mill was in operation again and paper was being sold with the new emblem.

The new Klippan Mill came into being at a very auspicious juncture. Denmark was very short of paper, and with only a couple of small mills operating within the country, practically all paper had to be imported. This ought to have been very good for Klippan's business, but the real take-off did not come until some years into the 1660s.

Krefting's lease expired in the summer of 1668. Curiously enough, neither the authorities nor the master paper-maker, Samuel Dröscher, appear to have realised the fact; instead, Dröscher was given the lease of the mill from the

28

summer of 1667, for a term of 15 years and in return for an annual dues totalling 600 silver dalers. He was entitled under the less to take what timber he needed from the Crown forests.

Klippan at this time was powered by two waterwheels. These drove 16 stampers each, making 32 altogether. In addition to the master himself, the papermill was operated by three journeymen.

Dröscher was a vigorous an enterprising man, and before long he had added to the facilities an abandoned ironworks "which could not be kept going". He also secured a life-term lease of the waterfall, at an annual charge of 30 silver dalers. Conversion of the ironworks into a papermill was completed in 1671, and a wheel and 16 stampers came into operation that very same year.

In about 1680 Samuel Dröscher made over the mill to his son and namesake. The lease was due to expire in 1686, but the yuounger Dröscher requested the Governor General for a life tenancy. This was granted and he was promised security of tenure so long as he kept the mill in good repair and paid his dues promptly. Later on he managed to extend the leasehold to include his children, "that they too, without let or hindrance from any other, might have possession of the mill and enjoyment of all the priviliges attaching to the same, for as long as they are ready and able to keep the said papermill in proper condition and to perform the duties and obligations incumbent upon them."

Klippan's watermark appears more and more frequently in documents and printed books at the end of the 1660s. The great jurist Samuel Pufendorf had both his *De jure naturae et gentium* (1672) and the shorter *De office hominis et civis* (1673) printed on HVC paper.

In the mid-1670s, Klippan acquired a competitor in Bishop Peder Winstrup, who had established a small papermill at Värplinge, outside Lund and did not balk at announding from the pulpits "that none might sell rags but to his, the Bishop's papermill".

More and more papermills were started in the provinces of Skåne and Småland, and there was fierce competition for the necessary raw material - rags. On top of this, trade suffered from the Skåne war of 1676-1679, which included a prolonged Danish occupation of the country round about Herrevadskloster.

These, however, were passing misfortunes. In 1679 peace returned, Bishop Winstrup died and Klippan recovered its position in the official documents of the Governor General's office.

A new slump came at the end of the 1690s, when the rag collectors were conscripted for military service. Rag supplies were cut off, the little papermill ground to a halt and before long it was quite unserviceable. The dams were destroyed by ice and cutbacks also occurred at the main mill. On Dröscher's death in 1714, his widow, Janeka, was left with a fairly dilapidated installation.

During the Age of Liberty in the 18th century, demand for paper in Sweden greatly outstripped supply. Due to the shortage of homeproduced paper, the mills could always be sure of a sale for their output. Failure to measure up to the quality of, say, Dutch paper was put down to a shortage of better-quality rags. In 1783 there was promulgated an "Ordinance on the collection of rags for the papermills of the realm", requjiring the master and mistress of every household in town and country alike to collect linen or woollen rags of the better sort as well as paper waste and trimmings. The papermills for their part were obliged to organise and finance the collection of rags.

Klippan had had its ups and downs in the two decades preceding the 1783 Rag Collection Ordinance. Janeka Dröscher carried on the paper-making business after Samuel's death in 1714, but output was apparently negligible. In 1722 the mill passed, by inheritance, to the son, master paper-maker Hans Dröscher.

By now the papermill was completely run down and a thoroughgoing restoration was needed before it could resume operations. Extensive repairs were carried out between 1722 och 1727. The lesser mill, for example, was completely rebuilt, while the larger one was given new foundations and stampers. New moulds were imported from Lübeck; attempt by Klippan to produce its own are immortalised by a number of incredibly deformed watermarks. A large copper cauldron was acquiured for boiling glue, and new felts were also purchased.

By 1742 the papermill was once more so dilapidated and "for the most part in such a state of ruin and decay that it can hardly be reinstated without a good deal of repair". Operations suffered a great deal from the numerous repairs and the shortage of capital. Output at this time was in the region of 2,000 reams annually.

Most of this paper was sold in Skåne, but printing paper was also supplied to other parts of Sweden. Karl Gustaf Berling, the Lund printer, was Klippan's biggest printing paper customer in the 1760s.

In 1772 Klippan was acquired by Abraham Tornérhjelm, who did his best to revive its flagging fortunes and improve local conditions. Among other things, he build a new bridge over the Rönne river. The changes and improvements which now take place ushered in a new period of prosperity, and by the mid-1790s there cannot have been more than three or four papermills in Sweden larger than Klippan.

In 1799 the mill was taken over by a civil servant, Anders Hellman, and a clerk, Carl Fredrik Kemner. Inspection revealed it to be "damaged by rot and impossible to repair, the locks for the waterwheels and in the canal as well as the water pumps and likewise all the bridges were found to be unserviceable, added to which the greater part of the timbering of the great dam across the River Röen is so badly damaged that it will have to be replaced or the dam lost".

Surviving documents and maps furnish a detailed picture of the mill facilities at the beginning of the 19th century. The northern part of the 36-metre-long, two-storey building erected in 1774 housed a wheel shaft rising the full height of the building. Water was taken from the canal and over the spillway through wooden channels to the waterwheel. The latter had a six-metre cross-section and was attached to a seven-metre-long oak shaft fitted at both ends with iron journals running on iron sockets. On the lower storey, next to the wheel-house, was the "roll-house" with the "rolls" or hollanders. The waterwheel was geared to two shafts, one of which drove three hollanders while the other powered a stamper and two water pumps. The wooden shafts of the rolls in the three hollanders had been replaced with iron shafts and journals. The stamping mill consisted of "eight stampers in their oaken frame, the rags being stamped in a trough of oak with, on top of the same, a stamping chest composed of deal boards and a lid". The roll house also included two stuff chests and "a strainer or two-inch deal boards, for cleaning the water for the roll chests". A primitive kind of waste water processing plant, in other words.

The bottom storey of the southern part of the building was occupied by the workshop, with its three brick furnaces. At two of these there were vats and copper "blowers". The third had an in-built copper cauldron of just over 400 litres capacity, for boiling glue leather. The workshop also contained four presses - three of iron, one of wood - a windlass and three stuff chests. Some thousands felts and ten pairs of moulds of various kinds were kept in a special store room. The upper floor housed a small sorting and packing room with three iron presses, while the attic was divided into three lofts for drying paper, "which in the lower room is hung up on slender poles of pine and spruce in three lines of tribles. Close to the ceiling in the lowest department and everywhere in the upper room, paper is hung on thin raffia lines."

A few years later Hellman made over his share of the property to Kemner, in return for 10,000 riksdalers. Kemner's aim of producing paper at Klippan which would bear comparison with the best foreign qualities, however, was difficult to achieve without further improvements. Various changes were therefore made to the equipment of the papermill down to 1808. A third vat was installed and new machinery and implements were procured, among them a mechanical cutter with two blades and two rag dusters. At the same time, structural enlargements and alterations were made, and the mill, which had caught fire in 1805, was refurbished.

These changes and improvements greatly augmented Klippan's output capacity. In 1792 the mill was able to turn out some 2,500 reams, but in 1803 it was producing nearly 7,000.

To improve his exploitation of the property, Kemner decided in 1808 to

31

augment his operating capital by forming a limited company, to which the property and paper-making business would be transferred. The share capital of 57,000 riksdalers was divided into 320 shares with a face value of 175 riksdalers each. Kemner retained one-third of the shares for himself. The subscription invitation held out the prospect of an annual profit of 24 riksdalers per share, "with all facilities operating at full capacity and barring accidents".

As a result of Kemner's acumen and keenness, Klippan developed into one of the foremost papermills in Sweden. In the course of 25 years he had turned a fairly dilapidated plant into a flourishing concern. When in 1824 he retired from managing the enterprise, he had at close quarters a man to whom he could confidently entrust the care of his life's work, namely his son-in-law, Sven Magnus Sunnerdahl. Sunnerdahl had studied commerce in his native Göteborg and been granted the liberty of the city, as a merchant, in 1810. In 1819 he moved to Helsingborg and married Hedvig Charlotta Kemner. As soon as he took over Klippan in 1825, he embarked on a vigorous programme of expansion and improvement. The beater room, rebuilt by Kemner in 1823, acquired a fourth hollander and the mill was turned into a papermill, with two hollanders and other accessories.

As a businessman, Sunnerdahl was indeed a worthy successor to his father-in-law, and he kept close track of the many inventions that were now being made in the paper industry in other countries. A mechanical rag cutter, for example, the first of its kind in Sweden, was purchased from abroad.

Sunnerdahl's farsightedness quickly bore fruit in the form of a substantial growth of output. Whereas between 1820 and 1824 production had averaged 9,330 reams annually, between 1827 and 1831 it averaged 12,503, peaking in 1829 at 13,882, the largest quantity ever produced at the handmade papermill. Klippan was now the biggest papermill in Sweden.

The growth of output was matched by a rise in quality. The rags were more completely macerated than before, the knots that had previously been so common were a thing of the past, and the paper displayed excellent purity. New grades came on the market - bank note paper, veilum paper, wallpaper, white and blue sugar paper etc.

In the space of a few years, Sunnerdahl's exertions "to expand the facilities and bring them to their utmost completion" had been so successful that he could justifiably claim that "production here, as regards both the quantity of paper and its generally acknowledged eminence in quality, undeniably surpasses every other papermill in the country".

Then, all of a sudden, this paramount position was in jeopardy. At Strandmøllen in 1829, Joh. Chr. Drewsen had erected the first paper-making machine in Scandinavia. Both the quantity and the quality of the paper sold from that mill bade fair to deprive Sunnerdahl of his entire Danish market. Sunnerdahl visited Strandmøllen to take a closer look at mechanical paper-making, and he realised that "if I cannot manufacture paper which will rival the foreign product in quality, then either my plant will run at a loss or else it will stand idle for a large part of the year". He lost no time in deciding: he would install similar machinery at Klippan.

But the acquisition of a paper-making machine and all the necessary alterations and extensions this involved was an enterprise which exceeded Sunnerdahl's financial resources. He therefore applied to the Manufacture Discount Fund for a loan of 15,000 riksdalers, and this was awarded him in November 1830. The following year a new building of granite rubble was erected to house the paper-making machine. The machine, imported from England, was 12.5 metres long, 1.93 metres wide and capable of producing a 1.35 - metre - wide web. It had five copper drying cylinders, above which a vapour hood rose nearly two metres above roof-level. The machine was actuated by a water wheel inside the engine house, and the water was taken through the room in a wooden channel.

Sweden's first paper-making machine started up on 20th June 1832. Klippan's machine-made paper output for the first six months was 6,677 reams or 44,335 kg. The following year output rose to 19,716 reams or 130,914 kg, climbing still further in succeeding years. This growing output called for a progressively larger supply of raw material, especially rags, and Sunnerdahl appointed special commissioners in Lund and Ystad.

Production mainly consisted of high-quality writing paper. The breakdown of output was roughly 60 per cent writing paper, 28 per cent printing paper and 12 per cent wrapping paper etc. During the closing years of the 1840s the mill also began producing lithographic printing paper and poster paper. Once the paper-making machine had got into its stride, output of the handmade paper mill rapidly diminished. The mill could still turn out 9,630 reams in 1832, but for the following year this had fallen to 1,723 reams. Production mainly comprised coarser grades of paper, especially wrapping paper for the firm's own use.

As regards the volume and quality of its output, Klippan at this time was the foremost papermill in Sweden, producing 20 per cent of all paper in the country. Compared with Sweden's other 88 papermills, Klippan's output was immense and impossible to compete with. The only papermills with any chance of survival were those which followed Klippan's example. Once this has been realised, paper-making machines were installed at a number of mills before the 1830s were over. In the Swedish Statistical Abstract for 1836, we read that the Klippan papermill in the County of Christianstad, belonging to Mr Sunnerdahl, had the largest output in the country, "as a result of the improved mechanical devices for

*Klippan PM 1. Sweden's first paper-making machine was installed in 1832 and remained in service until 1982. The machine is still to be seen at the Klippan mill. Photograph: Klippan Archives.*

this industry which he has introduced".

The following note is appended: "There are few branches of industry where such great improvements have been invented and introduced in recent times as in the manufacture of paper. A Mr Dickinson of Hertfordshire in England has invented a quite complicated but nonetheless inexpensive machinery with which paper is produced with the greatest accuracy, consistency and economy of price, far better than the handmade product. What used to take three weeks to complete in the production of paper is now accomplished in three minutes! A continuous fluid stream of pulp is converted, within the short distance of 15 ells, into an outstandingly beautiful paper which is dried, glazed and cut on all four sides, leaving it ready for immediate use."

Klippan, then, was a pioneer in many ways and the Swedish paper industry owes a manifold debt of gratitude to the farsighted, vigorous manufacturers associated with the trade's infancy.

Apart from the Klippan success story - and Klippan strictly speaking, began life in Denmark - Sweden's first papermill worth mentioning seems to have been established in 1612, when Gustavus Adolphus commissioned the German Arnold Schlodt to set up a papermill in Uppsala. This new industrial venture, then, coincided with Gustavus Adolphus' accession to the throne, by which time Sweden had been without a papermill for 40 years or more. The wretchedly small volume of paper imported, moreover, meant high prices, and so Gustavus Adolphus found himself compelled to do something about the matter.

Only two months after the death of his father, King Karl (Charles) IX, Gustavus Adolphus circularised his sheriffs, ordering them to make arrangement for the collection of rags for the new papermill in Uppsala. The letters patent explain that a lot of paper, "not all of which should be purchased so expensively from foreigners", was needed for the Chancery and other purposes. The papermill got through a great deal of linen and every year the sheriffs were to collect rags from farmsteads in the countryside. (Letters patent, Concerning wornout linen garments for the papermill in Uppsala, 8th January 1612.)

Building work on the papermill proceeded apace and it must already have been operating by the autumn of 1612. Eventually Schlodt abandoned his first mill, a converted hammer forge on the Fyrisån River, and had a papermill built at Sandvikshagen downstream of Uppsala, still on the Fyrisån River. One of the benefits of the new location was that paper could now easily be shipped out to Stockholm.

Schlodt was discontented, however, because of the shortage of pure water for the Uppsala mill and the inadequacies of the dam constructions at Sandvik, and so he migrated south in search of new pastures. Gabriel Oxenstierna's estate at Tyresö had several good waterfalls which Gustavus Adolphus himself had at one time thought of harnessing for Crown factories. The new facility, Uddby Papermill, was built, not later than 1621, by Arnold Schlodt, who presumably then paid an annual rent for it. Schlodt ran the mill until his death in 1644, and in 1649 Uddby Papermill was leased to Anders Dames who had married Master Arnold's widow.

Little was known concerning the fortunes of the Uppsala papermill after Schlodt had left it, until 1626, when it was taken over by another German, Hans Obenher. He must have been more of a businessman than the two journeymen who had taken over from Master Arnold. Output rivalled that of Uddby, although the Uppsala paper was generally inferior in quality. Owing to the bad water supply from the Fyrisån River downstream of Uppsala, the Provincial Governor, Johan de la Gardie, requested in the autumn of 1632 that a site for a new papermill be made available at Ekeby, still on the Fyrisån River but at the northern end of the Parish of Old Uppsala. Nothing came of this, however, and paper production continued in Uppsala.

In December 1641 Obenher requested the Uppsala University Senate to grant him "his customary liberty of a post horse while collecting rags in the country at large".

At the end of the 1640s Obenher decided to transfer the papermill to a spot with a more adequate water supply, namely Lötegårdsbro in Old Uppsala. The new facility was apparently under construction at the beginning of the 1650s, when Obenher requested a loan of 200 riksdalers for the purpose. The University Senate demurred, whereupon he went down to 100 dalers. In August 1652, reference is made to "the paper-maker's request for money to repair his factory". Sad to say, things apparently got the better of the unfortunate Obenher, and he died a pauper in 1656. Paper manufacturing in Uppsala continued, however, under the direction of other paper-makers. In 1637 Nils Månsson and Anders Mattson of Norrköping were granted a charter for a papermill in Fiskeby, and one of Sweden's best-known papermills, Lessebo, in southern Småland, was founded at the close of the century. In 1692 the Konga Hundred Court gave orders for the marking of oaks for Lessebo papermill, which suggests that the mill was under construction at this time. Papermills were very often set up to replace ironworks, and sometimes they formed part of a small industrial conglomerate, together with tilt-hammer forges, flour and groat mills and brick kilns.

Lessebo began manufacturing paper at a time when nothing had happened to advance the state of the art. The same painstaking, complicated methods were being used as when paper-making had first become known in Sweden, roughly a hundred years earlier. Rags were the only raw material. After they had been collected, they were first given a perfunctory cleaning and then graded according to purpose, fineness, whiteness etc. Next they were cut up and washed. In some places they were first dumped in great piles to ferment.

After fermentation, the rags were macerated in a stamping mill. This consisted of a large wooden block into which a number of holes - vat-holes - had been hollowed out. The rags were macerated under a steady stream of water in the stamping mill, a process taking about 12 hours. The pulp which resulted was collected in a half-stuff chest. After further maceration, this time for 24 hours, the pulp would be of sufficient fineness and was collected and transferred to the whole-stuff chest.

The pulp was now ready for forming, and the whole-stuff was transferred to the vat. There it was diluted with water into a thin slurry, which was kept constantly in motion. The vat was heated by means of a copper boiler and the paper was formed in the same way as today. The vat team used two moulds and a deckle. The vatman formed a sheet, let the water run off the mould, removed the deckle and passed the mould to the coucher, who deposited the sheet on the felting. He then returned the mould to the vatman, after which he placed a new felt over the newly formed sheet, assisted by the layboy. The entire procedure was then repeated until a post - about 150 sheets - had been made up. The water was then squeezed out and the sheets hung up to dry. Mostly they would be dried in the lofts above the papermills, but larger mills had special drying sheds. After drying the paper had to be sized, otherwise it could only be used for printing and wrapping purposes.

Lessebo at the end of the 17th century was quite an impressive establishment. It consisted of a two-storey factory building and, close by, "a cottage and other small buildings for the needs of the paper-maker". The ground floor of the factory building housed the stamping mill, vat and other devices for paper-making, while the upper storey was used for drying. The stamping mill had eight stamps and four hammers in each block. By comparison, the Klippan Mill in Skåne during the 1670s had 48 stampers.

The Lessebo Mill seems to have been founded to meet demand for cartridge paper, which had previously been imported. No production apparently materialised, however, until 1697, when 50 reams were delivered to the Navy in Karlskrona at a price of 7 silver dalers per ream. Governor Rudebeck, joint owner of the mill together with the Crown, requested "sole privilege of supplying cartridge paper to the Admiralty" when "producing as large a quantity as is annually needed and as good as comes out of Holland".

But cartridge paper never became a big item with the Lessebo Mill, nor was the Governor successful in his petition for customs exemption for Lessebo products. On the other hand, Lessebo does seem to have been manufacturing writing paper from the very beginning of its career. A watermark in the form of IB as a mirrored monogram surmounting two crossed twigs and crowned with an emblem has been discovered in paper dating from Jöran Branting's time.

This watermark, first known to occur in paper from 1695, was soon replaced by a new watermark consisting of the letters PR - Paul Rudebeck - in a mirrored monogram. Like Branting's cipher, it was positioned over two twigs laid crosswise, but this monogram was surmounted by a royal crown - a curious way of representing two half-owners of the mill, the Crown and Rudebeck.

In terms of quality, Lessebo paper at this time does not differ from the products of the other Swedish mills. In terms of fineness and consistency, they were all a good deal inferior to the Dutch and French paper in use.

A report to the Governor of Kronoberg County states that in 1748 Lessebo produced 90 reams of writing paper, including 50 reams of the coarser variety, 40 reams of printing paper, 200 reams of grey paper, 100 reams of tobacco paper,

*Left: Primitive rag processing at the Ösjöfors Mill. Right: Wooden water wheel at Ösjöfors.*
*Photographs: The Swedish Museum of Technology.*

10 reams of cartridge paper and 20 reams of ammunition cartridge - all of which adds up to mere 460 reams for the year. Lessebo, then, was one of the least significant papermills in the country.

In 1811, however, a new manager - Anders Wintzell - entered the business and personnel strength was raised to nine journeymen, two lads, one beater man and two girls. Colouring the paper made it also possible to use rag of inferior whiteness. That same year a start was also made to sizing the paper, to improve the quality and appearance of its surface, and the paper moulds began to be used with greater accuracy.

These improvements heavily increased the output of writing and printing paper, at the same time as production of other qualities was restricted. Output rose from 1,059 reams of writing and printing paper in 1807 to 2,480 reams in 1811.

By the end of the 1820s the paper-making machine patented by Louis Robert in 1799 had been introduced above all in England but in other countries as well. The first of these machines in Sweden was commissioned in 1832 by Sven Magnus Sunnerdahl at Klippan, which had been Sweden's biggest papermill since the mid-1820s. The paper manufactured there was superior in both quantity and quality to anything which the other papermills of Sweden were capable of producing. Klippan was a formidable rival.

Johan Lorens Aschan, the owner of Lessebo at that time, realised that, if his mill was to survive, he would have to mechanise production. Work began in 1833 on the new buildings required for the machinery. The hollander room consisted of four hollanders, built of four-inch deals and with beech rolls, each fitted with 48 knives. They were 3 metres long, 1.4 metres wide and 0.6 metre high. They were estimated to hold about 40 kg of pulp at 3 per cent concentration. Otherwise the commonest arrangement at this time was for the hollander to hold about 50 kg pulp. Motive power was obtained from a water wheel 5 metres in diameter and 2.4 metres wide. This part of the mill, known as the South Wing, was completed in the autumn of 1835. The North Wing was fitted out for finishing operations. In addition, a vaulted room was reserved for the boiler which would supply steam to the drying cylinders of the paper-making machine. The latter had been ordered from Bryan Donkin & Co. of London, at that time the foremost manufacturers of such machinery. F.o.b. London, the machine cost 26,410 riksdalers. When it finally arrived at the Karlshamn docks, there was the difficult business of getting it to Lessebo. Two sledges broke under the weight, for example. But by about Christmas 1835 the machinery was on site, and in March 1836 Mr Thomas Seamer arrived from the factory to direct installation work. Sweden's second paper-making machine started up on 28th May 1836. It was 12 metres long and 1.97 metres wide between the uprights. The 8.46-metre-long wire of brass gauze was 1.48 metres wide, and so the working width of the machine must have been about 1.3 metres.

The number of papermills grew steadily during the 18th century, and during the centuries that followed these mills came to mean a great deal to the backward countryside. Every mill gave one or more families a livelihood, and local boys became apprentices and journeymen. After journeying in the old tradition, they themselves then became master paper-makers. Another profession equally essentially to the survival of the mills - the rag collectors - was also trained in close conjunction with the paper industry. Insuring an adequate supply of raw material was a constant worry to the paper-makers. Since that material consisted entirely of rags, the authorities had to give orders for rags to be collected for the benefit of paper factories. For at that time most old rags were re-used. From several counties reports have come down to us of rags being used to make bedcovers. Old fabrics were unpicked and used for weft, broadcloth and woollen fabrics being so expensive.

Rags, then, had an economic value which ought to merit compensation. There was no reason why the general public should voluntarily submit to all the inconvenience which rag collection entailed, but here as in many other fields, appeals to their sense of patriotic duty were combined with ordinances, edicts and penalties.

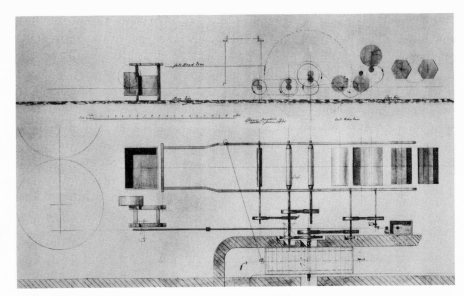

*Drawing of the paper machine supplied to Lessebo by Bryan Donkin & Co. on 1st June
1835 (excluding a couple of structural details). Lessebo Archives.*

Several provincial governors recommended that each papermill should be
allotted a rag collection district. Opinions varied, however, concerning the
actual process of collection. Certain governors felt that the rags ought to be
delivered to the papermill commissioners, whose duties would thus be confined
to receiving the deliveries and paying for them. Others felt that the papermills
ought to employ rag collectors who would have the menial task of going round
collecting rags from individual households.

The basic principle underlying the new regulations was that households were
to be obliged by law to salvage both fine and coarse linen and waste paper. The
papermills, however, were many in number, and no matter how zealous the rag
collectors and rag commissioners might be, there were simply not enough rags to
go round. As more and more members of the community learned to read and
write, paper consumption skyrocketed.

It was the same problem all over Europe, and everywhere experiments were
in progress to try and find a substitute for rags as they became more and more
difficult to come by. Already in 1684, Edward Lloyd in England had suggested
manufacturing paper from Linum asbestinum, of which there were generous
supplies available in the Isle of Anglesey. In 1716 another idea could be studied
in a small journal entitled Essays, for the month of December 1716, to be
continued monthly, by a society of Gentlemen. For the benefit of the People of

40

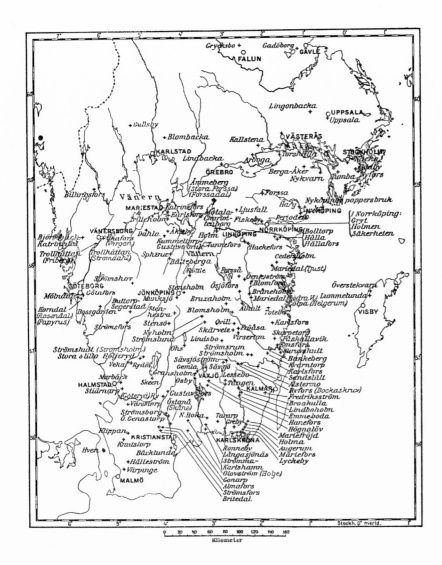

*Map over known paper mills in southern Sweden in mid 19th century. From Ambrosiani 2.*

England. This society had the idea of making paper from raw hemp, without the need of spinning and weaving it.

The first person to hit on the idea of making paper from wood was the Frenchman René Antoine Ferchault de Réaumur (1683-1757). An eminent naturalist and physicist, he had concluded, from his observations of wasps and their nest-building, that the wood fibres they used could also be turned into

41

paper. In 1719 he presented his findings and theories to the French Royal Academy: "The wasps teach us that paper can be made without the use of rags and linen. They seem to invite us to try whether we cannot make fine and good paper from the use of certain woods. This study should not be neglected, for it is, I daresay, important. The rags from which we make our paper are not an economical material and every paper-maker knows that this substance is becoming rare." Although there is no evidence of Réaumur actually having succeeded in making paper from wood, he was unquestionably the first scientist to put forward the idea.

In 1741 Jean Etienne Guettard, a member of Académie des Sciences, and physician to the Duke of Orléans, wrote several articles about making paper from conferva (swamp moss) instead of rags.

In the 1740s, Jonas Stakel, master paper-maker at Östanå Papermill in the Swedish province of Hälsingland, experimented with wood chips and leaves. His were the first attempts to be made in Sweden at finding a substitute for pulped rags. The diary of the Royal Academy of Sciences for July-September 1751 records that Stakel had submitted specimens of his paper for assessment: "H. Stakel, Manager of Östanå Papermill in Hälsingland, has presented specimens of a kind of grey paper made solely from leaves, with lime water and other additives, the revelation of which he reserves for himself, without the slightest quantity of rags being present. Similarly, another kind of paper, which in colour and firmness appears to bear a close enough resemblance to cartridge paper, made in the same way as the aforementioned, the sole difference being that a little sawdust is included instead of leaves."

One of the most interesting and unusual experiments in the search for substitutes for rag was performed by Jacob Christian Schäffer (1718-1790), a Regensburg clergyman who was a great authority on botany and natural history. His interest in the Bavarian flora led him to investigate the possibilities of finding new materials for making paper.

In 1765 he began a series of experiments with vegetable fibres which he recorded in a six-volume treatise, completed in 1771. He himself says that it was not his intention to make the best possible paper but to show that vegetable fibres could be used for paper-making. Curious to note, one of the first paper specimens in his treatise is made from pulp taken from wasps nests - the wasp, as Réaumur had already pointed out, being the world's first paper-maker.

Schäffer's experiments extended over a period of eight years, and all the material he used came from his local surroundings, e.g. potatoes, nettles and straw. In 1801 Matthias Koops of James Street, Westminster, opened a papermill for making paper out of straw. This was the first papermill in Europe to use a raw material other than linen and cotton. Unfortunately the venture did

*Réaumur finds food for thought.*

not work out as well as Mr Koops had anticipated, and only two years later his company went bankrupt.

The search for materials to be used instead of rags continued well into the 19th century but meantime efforts were being made to improve the yield on the raw material available. Writing paper and better-quality white paper could only be made with fine, white linen rags, but dyed, coarse linen was good enough for coarser grades. To produce absolutely white paper, however, the white linen rags, after they had been turned into half-stuff, had to be bleached in the sun and water poured over it. To avoid the time-consuming bleaching procedure and the yellowish-brown tinge of paper made from unbleached rags, the paper would be dyed a pale blue. This way, superior grades of paper could also be made from coarse linen rags. At the end of the 18th century, bleaching methods began to be introduced which made it easier than ever to utilise the rag supply available.

After K.W. Schéele had successfully produced chlorine in 1774, the French chemist C.L. Berthollet demonstrated, in 1785, the capacity of chlorine gas for bleaching and destroying vegetable dyes. The usefulness of chlorine as a bleaching agent was still further augmented by the English chemist Smithson Tennant's production of chloride of lime. Both these procedures were quickly

43

*Östanå Papermill, founded in 1664. Artist's reconstruction by Carl D. Svensson, based on the 1771 inspection report and other sources.*

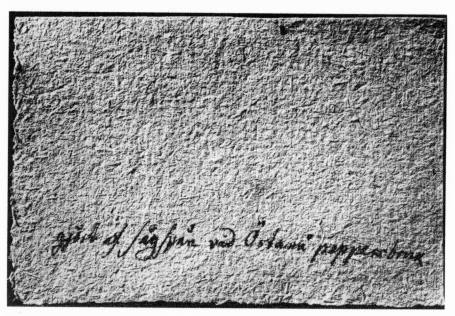

*Stakel's specimen of paper made from sawdust. Photograph: Swedish Museum of Technology.*

introduced by the papermills, and dyed linen rags could now also be used for producing high quality paper. This paper too, however, had to be tinted blue, because otherwise the chlorine-bleached pulp gave it a faint tinge of yellow. The bluish colour, therefore, is typical of most European and American paper during the decades round about the end of the 18th century. Most absolutely white paper at this time came from Holland.

Right down to the beginning of the 19th century, when the paper-making machine was just about fully evolved, all paper was being formed by hand. Consequently sheet sizes were restricted to the format of the paper-maker's mould, which in turn restricted the ability of printers to employ larger formats. All through the history of printing, both in the Orient and in the western world, paper has been the decisive factor.

As we saw earlier, there was a perpetual shortage of raw materials for paper-making; the supply of linen and cotton rags was nothing like equal to the demand. This shortage attracted the attention of scientists to the problem, including them too look for new fibres which could be pulped for paper. This quest for new materials is believed to have inspired Nicholas-Louis Robert with the idea of his paper-making machine, although the constant bickering between the members of the vat team was a contributory cause. Born on 2nd December 1761, Robert was frail and sickly throughout his formative years. After attempting a military career and in fact taking part in a battle against the British, he returned to Paris at the age of 28 and entered the publishing firm of Didot. Tiring of his clerk's existence after a few years, he requested a transfer to Didot's papermill in Essones. This brought him into closer touch with hand-made paper-makers, and he was appalled by their lack of discipline. He had not been at the mill for many months before be began making plans for a machine which would simplify the production of paper and dispense with the chronically quarrelsome couchers, vatmen and charlatans in the industry.

Assisted by Didot, Robert designed a small-scale paper machine and carried out experiments. The first machine, however, was a failure, and if Didot had not been so enthusiastic Robert would probably have left it at that. Didot arranged for Robert to be transferred to a flour mill not far away, so that he could work undisturbed on his invention. Several highly trained technicians were placed at his disposal. Under Robert's direction the technicians continued working on the design, but with only limited success. Even so, Robert was convinced that he was moving in the right direction and that a paper machine would have to be designed in the manner he had devised.

In the end he successfully produced a few sheets which both he and Didot considered satisfactory, and Didot urged him to patent his invention. Robert took his advice and on 9th September 1798 formally applied for a patent for the

machine, which was capable of producing paper in an endless web.

The authorities realised the immensely important part that Robert's machine was destined to play in the paper industry, and they immediately sent a member of Conservatoire des Arts et Métiers to Essones to help in the preparation of drawings. Robert's own description (here somewhat abridged) reads: "At the end of the cloth wire extending on the vat there is a flywheel, or cylinder, fitted with little buckets which plunge into the paper stock, or liquid pulp. This cylinder, by its rapid movement, raises the material and throws it into a shallow reservoir in the interior of the head, and from there the stock is poured, without interruption, onto the endless wire cloth. As the material settles on the cloth it receives a side-to-side movement, causing the fibres to felt together. The water drains off the wire and back into the vat. A crank turns the machine and causes the wire cloth to advance, the sheet of newly formed paper finally running under a felt-covered roller. When the paper leaves the first felt roller it is no longer saturated with water, but can be removed from the machine, just as a sheet of hand-made paper is taken from the felting after pressing in a press."

It should be remembered that all printing at this time was done by hand and that there was no need for an endless web of paper. Accordingly, the paper was removed from the machine one sheet at a time. Later on this paper was to bring about a new revolution, this time in printing, because the endless web made possible the introduction of the rotary press.

After Didot and Robert had quarrelled about the future of the machine, Robert sold the patent to Didot for 25,000 francs, payable by instalments.

No payment was received and Robert took back the patent in 1801. Meanwhile Didot had told his brother-in-law in London, John Gamble, about the design and suggested that he try and build a similar machine in England. As a result of Gamble's efforts, two stationers, Henry and Sealy Fourdrinier, became interested in developing the new machine, and, through Gamble, a talented mechanic, Bryan Donkin, was engaged to construct a paper machine modelled on Robert's. Several new patents were taken out, and in 1803 Donkin had a machine ready which could turn out quite decent paper. This machine had cost the Fourdrinier brothers about $60,000, but like Robert, they never made anything out of it. To this day the design carries their name, but that is the only recognition they ever received for their perfection of the machine.

During the almost 200 years that have elapsed since Robert began developing his ideas about "endless paper", the Fourdrinier machine has almost completely eliminated the making of paper by hand. The latter is now almost exclusively an

*Previous page: Robert's paper machine, from the patent application in 1798.*

MAKING
PAPER
antiquarian pursuit, or else paper is made this way when almost infinite durability is needed. The Lessebo Mill is the only plant in Sweden producing hand-made paper on a commercial basis, and the hand-made papermill there offers guided tours for visitors during the summer months. At Tumba Bruk there is a paper museum which opens one weekend every month, and the old Ösjöfors Mill near Vimmerby is open to interested visitors during the summer season. Ösjöfors belongs to the Swedish Museum of Technology in Stockholm, and during the summer it is operated by students from the National College of Fine Arts.

# THE PAPER MAKER

**Georg Anzelius recalls his life with Tumba Bruk**

By the mid-18th century, the bank notes issued by the Bank of Sweden had come to be widely imitated, even though each one of them proclaimed: "Any person imitating this note shall be hanged". The Bank Commissioners became increasingly concerned to put a stop to the forgeries. The paper used until then for the production of bank notes had been purchased from abroad, owing to lack of knowledge and experience at home. The first Swedish papermill was founded at the beginning of the 17th century, and in the first half of the 18th century papermills in Sweden were still few and far between.

The Commissioners believed that the forgers could be checkmated by manufacturing bank note paper in Sweden with certain secret devices incorporated. In a letter of 14th June 1755, they expressed their belief that forgeries could be prevented by "having a proprietary paper marked with certain special characters, the composition of which should not be readily identifiable, and which should be different from those of all other paper. Several designs are now under consideration, and it will be further improved and prosecuted. Production of the said paper requires a carefully preserved secrecy, which is not to be expected if the Bank should resort to any of the papermills established within the country, which moreover are by no means sufficient to furnish the realm with an adequate supply."

For this reason, the Commissioners requested authority to purchase Tumba Farm, at that time the property of Under-Secretary of State Edward Carleson. Their request was granted and Tumba acquired for the Bank on 29th July 1755. Bank Commissioner J. Teuchler became the first manager of the papermill. He was said to "have given satisfactory proof not only of diligence and skill and of dependable fidelity and zeal in the best interests of the Bank, but also of an understanding of matters pertaining to paper-making, and especially of such inventions as will serve to make a new kind of paper, being distinctive and not readily imitable".

One of Teuchler's first problems was getting hold of skilled workers. None were to be had within the country, and so they had to be recruited from abroad, preferably from Holland, which in the mid-18th century was the most eminent paper-making country. There was more to it than just calling in experts, however. The guild of paper-makers kept close tabs on its members and their secrets, and there were severe penalties for anybody attempting to carry Dutch industrial secrets out of the country.

Ther så händer, at någon hädanefter, ehwad han är i Banquens tienst, eller icke, på et eller annat sätt, förfaljkar, förändrar, eller eftergiör en Banquens Sedel, ware sig Låne-Banco-Attest, Låne-Banquens Assignation på Wäxel-Banquen, eller Låne-Banquens Bokhålleri-Sedel til Casseuren, eller Wäxel-Banquens Transport eller Cassa-Sedel, til större eller mindre Summa, antingen han thermed tilbringar sig penningar och waror, eller icke; Så skal en sådan, utan alt skonsmål, eller anseende til Personen, Stånd, eller wilkor, mista åra och lif, samt warda hängd.

För öfrigit är härmed Wår nådige wilje och Befallning, thet wederbörande domhafwande skola, utan uppehåll, alla sådane mål afgiöra, så at missgiärningsmannen måtte skyndsamligen blifwa afstraffad. Thet alle, som wederbör, hafwe sig hörsammeligen at efterrätta. Til yttermera wißo hafwe Wi thetta med Egen Hand underskrifwit, och med Wårt Kongl. Sigill bekräfta låtit. Stockholm i Råd-Cammaren then 2. Maji. 1747.

# FRIEDRICH.

## (L. S.)

*"… shall be hanged." The penalty for forging bank notes.*

The Bank Commissioners approached the Office of Manufacture, and the Swedish Minister to the Netherlands set about recruiting vatmen and rag-pickers. This was a very difficult undertaking indeed, due to the great vigilance and jealousy of the Dutch, but in the end, through the good offices of an Amsterdam merchant, the Minister managed to secure an agreement with two workers.

Preparations were made with the utmost secrecy, but the plan for bringing the two Dutchmen to Sweden was discovered at the last minute. Probably it was Jan Mulder, one of two brothers, who, unable to keep things under his hat, began boasting about his future career as a master paper-maker in Sweden. He was arrested and sentenced to 16 years' imprisonment, but he died after only a few months. The other brother, Erasmus Mulder, went into hiding in the countryside and eventually made his way to Germany and from there to Sweden. He was put on the Tumba strength in 1758.

Erasmus Mulder was the first master paper-maker at Tumba, but before long he was joined by his brother Casper and then, eventually, by other members of his family. As is also shown by the subsequent history of the Tumba Papermill, it was usual for several members of the same family to be employed at one mill.

In addition to various perquisites, Erasmus Mulder was paid the equivalent of 1,500 Swedish copper dalers in Dutch guilders. His contract exempted him from "all contributions and dues to the Crown" and he would not have to "concern himself with anything but, solely and exclusively, conducting the manufacture of paper as profitably as possible". In addition he was to make sure that tools and implements were "constructed and made in the best Dutch fashion".

The Mulders, a skilled family, occupied important positions at Tumba for many years. Erasmus was manager, Casper master paper-maker, Thomas first-packer, Cornelius second-packer and Johan sizer. The older members of the family seem to have been at pains to preserve the secrets of the art, but the younger ones both eventually took apprentices, among them Samuel Knauer who became a master paper-maker in 1799 and retired in 1825 after 60 years with the firm.

Eventually there developed the good spirit of tradition and practice which constituted the real secret of this old institution. Even if people supposed in 1755 that the art of making bank note paper required secrets and "peculiar inventions", this was not really the case. The art, or the secret, of it lies in the meticulous practice of a good method and the maintenance of a firm tradition from one year and generation to another.

Following Teuchler's transfer to another post within the bank, the printer Peter Momma was appointed to succeed him. Momma was an industrious, worthy man who knew good work when he saw it. What was more, he had spent

4 years studying and working in Holland, which had taught him a lot about Dutch printing and, probably, about Dutch paper-making as well.

Momma's first task at Tumba was to perfect the facilities. He took a close interest in all building enterprises, right down to the smallest details. Tumba did not have an adequate water supply, and a deeper canal was built. This was completed in the autumn of 1761. The channel between Lakes Alby and Tullinge was dredged so that rags could be shipped from Stockholm to the papermill, and Tumba took delivery of the first waterborne consignment on 28th June 1763. To begin with, rags were stored on rented premises in Stockholm, but in 1762 the Bank Commissioners purchased a property at Skinnarviken, complete with a building for use as a rag store. In 1758 the Commissioners had resolved that the workforce at the papermill was to consist of 1 beater man and his assistants, 4 rag-pickers, 3 workers at the vat, 1 sizer, who was also to tend the bundle presses, 2 workers for laying up the sized paper, 3 workers to sort the paper, 1 carpenter and 1 mould-maker.

In the autumn of 1759 the workforce thus resolved on was complete but Momma had a great deal of trouble with his mould-maker. The previous master craftsman, Runblad, had indeed made a number of moulds which were in use at the mill, but he had thrown up this job because he was unable to get on with other workers. And besides, he was not felt to have the necessary skill for making the figures needed for properly shaped, elegant watermarks and fitting them to the wire gauze. An engraver employed by the Bank, called Adam Renander, was entrusted during the spring of 1759 with the task of preparing silver figures for a watermark, to be attached to the paper-mould for four bank note forms. Momma was satisfied with his work. That autumn Momma recommended that a Stockholm master carpenter, Anders Bohman, who had already delivered moulds to Tumba, should be appointed mould-maker. Bohman had been abroad and studied mould-making, and among other things he had constructed two pairs of divided moulds, a device not previously seen in Sweden. Bohman was given the job, added to which he received an incentive payment for instructing an apprentice mould-maker.

In June 1759 Tumba delivered paper for the new bank notes. The design had been prepared by the medal engraver Carl-Johan Wikman, and the type-founder Matthias Holmerus. The new notes, in denominations of 9 and 6 copper dalers, carried the following warning in small print: "Any person imitating this note shall be hanged. But any person demonstrably discovering the imitator shall receive a reward of 40,000 copper dalers, pursuant to His Majesty's gracious proclamation of 20th December 1754."

Momma always took good care of his employees at the mills, a policy which did not always find favour with his superiors in Stockholm. The Bank

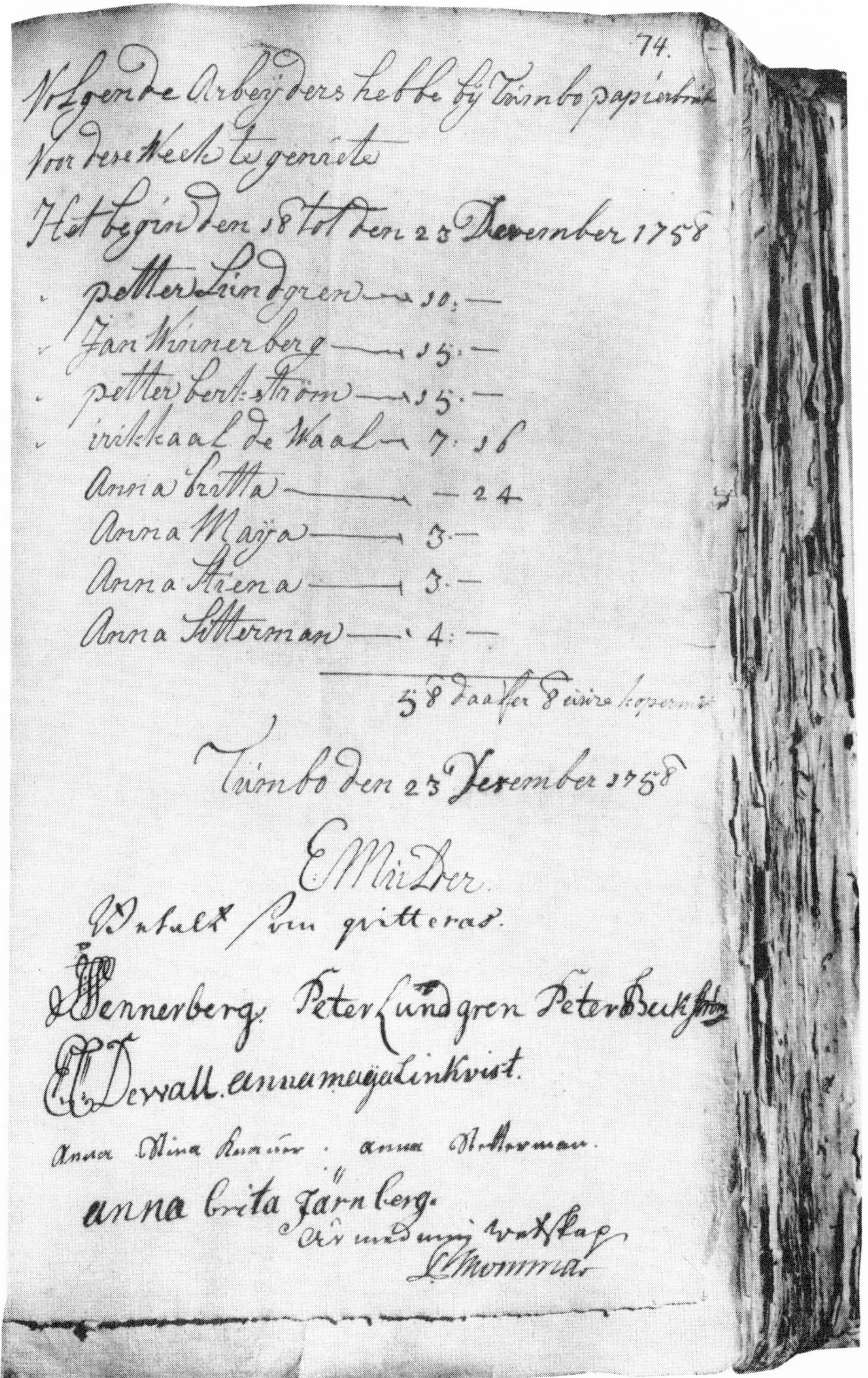

Volgende Arbeÿders hebbe bÿ Tumbo papierbruk

Vor dere Weck te genäte

Het begin den 18 tot den 23 December 1758

petter Lindgren ⸺ 10:—
Jan Winnerberg ⸺ 15:—
petter berk ström ⸺ 15:—
irikkaal de Waal ⸺ 7: 16
Anna britta ⸺ — 24
Anna Majja ⸺ 3:—
Anna Stiena ⸺ 3:—
Anna Sötterman ⸺ 4:—

_____
58 daaler d eire kopermÿt

Tumbo den 23 December 1758

E Wilder.

V. Intalt für quitteras.

J Wennerberg. Peter Lundgren Peter Bukström

C: Dewall. anna magdalinkvist.

Anna Stina Runür. anna Sötterman.

anna brita Jarnberg.
Är med mÿ vakttag
S Mommar

*Wage list, 23rd December 1758.*

Commissioners made certain critical remarks about the mills, with reference to the 1768 accounts. Among other things they noted that the workers were paid weekly wages all the year round. This should be rectified, so that no wages were paid for high days and holidays. It was also remarked that the wood fuel needed by the workers in their homes was being cut at the mill's expense. The Commissioners would prefer the workers to be given uncut wood.

Momma replied that his "purpose had ever been to make the mill-workers' conditions as good as possible without harming the mill, so that they would work with a will and take it as a sensible punishment if any were to be dismissed for misconduct".

There were constant differences of opinion between Momma and the other Bank Commissioners, partly perhaps because Momma's health had begun to fail. He had not had any time off between 1758 and 1770, and in addition to Tumba he had his own printing works in Stockholm to think about. In the spring of 1770 he requested leave of absence to take the waters at Medevi. Unfortunately this uncommonly active man did not recover his health but died on 17th March 1772.

According to Nils G. Wollin, Momma had wanted his grave to be marked with an iron slab bearing the following inscription:

Peter Momma

Bank Commissioner and Director of the Royal Printing Office

Born 21st April 1711. Died

He founded Tumba and Harg Papermills

And introduced the true art of paper-making into this country

As well as procuring type foundries for the printing works.

This versatile and immensely energetic man never seems to have been accorded a memorial acknowledging his life's work. Although several people have vied with him for the honour of establishing Tumba Papermill in the true Dutch fashion, it is mostly to his credit that Tumba did so well during its first decade.

By the beginning of the 19th century, the mill's output had assumed such proportions that it was producing all the paper needed by the Bank of Sweden, the National Debt Office and the Charter Seal Office. Examining old Tumba paper from this period, one finds that the product was of good quality.

Procurement of rags was the big problem, as with all other papermills. In 1803 this led the Bank Commissioners to offer a prize of 200 riksdalers to anybody "indicating the best and truly least expensive manner in which the collection of rags can be assured and most easily conducted". At the 1812 Riksdag in Örebro, the Bank Commissioners proposed "that the sextons in the parish ought on suitable occasions, e.g. after the end of divine service, collect whatever rags the

*Top left: New bank note proof, 1759. Top right: "The "Bancohuset" watermark. Bottom: Early watermarks. Tumba Bruk Archives.*

parishioners could deliver, and then dispatch them to the mill's warehouse in Stockholm".

The rags were conveyed by water to Stockholm, where Tumba Papermill had two warehouses: one at Skinnarviken, for rags coming from the region of Lake Mälaren, and one at Ladugårdslandet for those coming from the other direction. The latter building also housed part of the bank's type foundry.

A new papermill was completed in 1815. The forming room was equipped with Brahma water presses, the first hydraulic presses erected in Sweden. These had been delivered by Sam Owen's workshops in Stockholm. Major Mechanic Blom, supervising the building operations, was so attracted by the new presses that, in his opinion, Owen should have received a special reward for his "new invention of carrying out pressing by hydraulic power, as being, at least in this country, a new invention and very economical of labour".

Midsummer Eve, 1829 was a tragic day in the history of the mill. The Upper Mill was completely destroyed by fire. Apart from the building and its machinery, the fire also destroyed all the felts and moulds not in use that day, including 59 pairs of moulds for 45 different types of paper and 460 reams of paper which had been hung up to dry. A subsequent investigation revealed that the fire had been due to a faulty flue in one of the drying lofts.

Fortunately the Lower Mill could maintain the necessary output while the other one was rebuilt. The new building was very solidly built and roofed with copper. It housed a forming room with two vats and a calender room, a bleaching room, a separating room, a drying room etc.

The Swedish Statistical Abstract for 1836 records that there were now 91 papermills in Sweden. Their combined output was 178,846 reams of paper, 6,167 lb sheathing paper, 111,518 sheets of pressing and roofing paper, and 2,552,600 forms for the new bank notes. All this was valued at 578,220 riksdalers. That same year 30,983 reams of various kinds of paper were exported and 2,182 reams imported.

Tumba Bruk, then, was responsible for the production of bank notes. Experiments with hand-moulded, multi-ply bank note paper had already begun at the end of the 1820s. In November 1825, Professor G.E. Pasch wrote a letter to the Bank Commissioners describing his visit to papermills in France, among them St Marcel outside Lyon. Here he had learned the procedure for making paper from straw. The difficulty involved in bleaching the straw, however, made this paper brittle and expensive. Another material which seemed easier to use was the waste and shives resulting from the hackling and braking of flax and hemp. This could be used to make a thin, tough paper on which the French printed their bank notes. The secret of the paper's composition, however, was jealously guarded by the French authorities. Nevertheless, Pasch contrived to see the various manufacturing processes and he actually brought specimens back with him to Sweden.

After his return home, Pasch expressed himself willing to begin experimenting on a small scale with the methods he had studied on his journey. He was already able to announce in 1827 that he had successfully produced four different grades of paper at Tumba Bruk, viz simple paper of plain hemp and pulp, double, two-ply paper, between the layers of which figures could be inserted as desired, a simple paper made from a mixture of hemp and rag pulp, and double paper of the same kind. Pasch considered his paper as good as the French varieties and believed that it could be manufactured at Tumba.

Jonas Bagge, Second Master at the Falun School of Mining, was made Pasch's assistant in 1833. He gave the following description of the working procedure evolved experimentally for forming the two-ply paper:

"In addition to the ordinary vat containing the stock for the outer layers, there is one more with the intermediate-layer stock, tended by a special vatman. Two moulds are operated for the outer layers, but for the intermediate layer only one, namely that containing the watermark.

"The team consists of two vatmen, a coucher and three boys, one of whom helps to guide the form during settling while the other two act as layboys, in addition to rinsing and wetting felts.

"Couching takes place on a wet felt, starting with an outer layer and continuing with the intermediate layer, on which the other outer layer is placed, whereupon the sheets thus formed are lifted away on their felt and put in a separate pile. To keep all three layers superimposed on one another, the moulds are fitted with spikes which engage in blocks attached to the asp. These blocks have hinges so that they can be lifted up every time the felt is removed or inserted. When the post is complete the water is squeezed out in the usual way. The felt post is then covered over, and a moistened and sized paper of inferior but strong rags is laid over every sheet of bank note paper.

"Meantime bank note paper is being formed and couched on another felt post and when this too is ready for pressing, the previously mentioned post, with interleaving paper between each felt, is inserted in the press at the same time. During pressing, these two posts are separated by a loose pressing board which, during the operation, is pushed up to the top of the press and retained there by means of a support at each end. An ordinary press plank is put on the top post, and between this and the top of the press is the wet interleaving paper, which will then be used for the bottom post. Everything is now pressed at once. The water is squeezed out of the bottom post, in the middle post each sheet of bank note paper is pressed against its interleaving sheet, and the interleaving paper above is pressed smooth and drained of superfluous water. The bank note paper pressed onto the interleaving paper is now put into reams and work is repeated on the felt post thus emptied, while the other one is provided with interleaving paper. Work thus proceeds on two felt posts, one of them being wrapped and provided with interleaving, while couching takes place on the other. A boy is occupied rinsing the felts clean and putting them in order for the coucher. Should it be observed during couching that any sheet is defective, this will be rinsed away at once.

"After couching the paper is squeezed dry towards evening, after forming is over, and it is then parted by the drying shed workers. During this latter parting operation, two sheets of bank note paper are always turned face to face, and these, together with their interleaving paper make up a spur, in which the bank note paper is always covered by the interleaving paper. It is pressed once more, very gently, and then hung up.

"Work on the simple notes also takes place with two felt posts, but this time the felts do not have to be wetted, so long as they are otherwise of good quality. Otherwise couching follows the same procedure as with ordinary rag paper, although the technique is somewhat different."

I can add to this account that the boy referred to resembles the layboy whose task was to rinse the felts in water. After doing this he would hand over about 25 or 30 at a time to the vatman, who together with the coucher would put them up on the bridge.

Bagge's statement that the multi-ply bank note paper was parted does not tally with my experience, but on the other hand, the simpler grades assembled in reams were re-pressed before being hung up to dry.

After the new mill, in Neo-Classical style, had been completed and opened in 1833, there were only minor changes. An extension was built in 1845 to house two bleaching chests of refractory brick and an apparatus for producing chlorine gas. This extension was converted in 1860 into a water reservoir with a sand filter through which water from the river was passed before being put into the process.

Water shortages frequently resulted in stoppages, and Bagge had therefore suggested that a steam engine ought to be purchased for the mill. The Bank Commissioners then appointed Deputy Director General A.E. von Sydow to investigate the water supply question. He suggested that a dam be constructed about 40 ells above the Upper Mill and a channel constructed from that point. The Bank Commissioners decided in favour of an iron channel, and this was delivered and installed in 1846. On the other hand, Bagge never had the pleasure of seeing a steam engine at the mill.

By the mid-19th century the Upper Mill was in such a state that it would have to be rebuilt. Both the masonry and the woodwork had been so badly attacked by damp over the years, due to inadequate ventilation, as to endanger the health

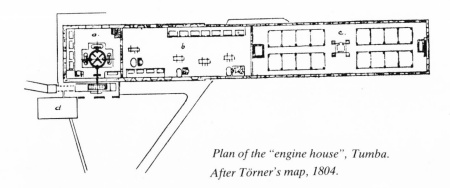

*Plan of the "engine house", Tumba.*
*After Törner's map, 1804.*

and lives of the workers. A scheme for giving Tumba such spacious facilities that it would be able to deliver all the paper which the State required was rejected as being too expensive. Instead a committee of inquiry came to the conclusion that the Lower Mill should be expanded and altered and production of bank note paper and safety paper transferred there.

The Upper Mill was repaired and converted into a rag store, a cutting room and a bleaching house. The portion of the building which had previously housed the boiler was turned into a bathhouse having two rooms and a bathtub in each - a remarkably hygienic arrangement for the time. One bathroom was for the clerks of the mill and the other for the workers.

Even today, the old building with its green copper roof and purity of line makes a venerable impression.

The Lower Mill was ravaged by fire in January 1875. The bitter cold - it was more than 30 degrees below zero - impeded the fire-fighting operations, because the water turned to ice in the hoses. Eventually the fire was brought under control, but by then it had caused such extensive damage that the Lower Mill had to be rebuilt. Before this was done, however, the Bank Commissioners instructed the manager to find out what he could about more modern machinery at a couple of papermills constructed recently in Northern Germany. On his return a scheme was drawn up for the new building. This provided for the mill to be powered by steam and the steam engine came from the Bolinder works in Stockholm. Some of the other machinery had been bought at an auction at Nykvarn Papermill, which had gone bankrupt at about that time, and the new building was completed in 1877.

A law passed on 12th May 1897 gave the Bank of Sweden a monopoly of all bank note emissions. The Bank Commissioners ordered an investigation concerning the increased demand for bank notes, and they also inquired about

*The packing room in the 1890s. Tumba Bruk archives.*

*Drying hand-made paper in the 1890s. Tumba Bruk archives.*

the arrangements needed at the mill to ensure an adequate supply of paper for all the new bank notes. The manager took the view that the mill could be extended 14 metres further east to make room for the two new double vats and their appurtenant hydraulic presses, vats for the moulding stock and so on. The beater room could be enlarged, so as to make room for another half- and wholestuff beater.

Work on the foundations began already in the autumn of 1897, and the new building was finished in 1899. At intervals of a few years, improvements and minor alterations were then made, with the mill operating to full capacity.

It was at this time that I first saw the light of day. I was born on 2nd September 1902, the sixth son of Anders Anzelius and Eva Myrman. In 1912 the family moved into a flat consisting of two rooms and a kitchen. With it went a bachelor's room, which was shared by two of my brothers. So when I was 10 years old there were eight of us living in this two-room flat, but we children felt we were very well off indeed. The whole family helped to tend the tiny garden which was also included.

In view of the housing situation for manual and white-collar workers at the end of the 19th century, I am sure we had reason to be glad. A few decades earlier, whole families had often been accommodated in single rooms which were partitioned off with a curtain, so as to give the impression of a living room and kitchen. Several of the kitchens had only open fire-places, so that if somebody wanted an iron stove installed, it had to be at his own expense and with the landlord's permission.

Eventually I began attending the infant school of the mill. My teacher was called Ada Molin and she also taught handicraft to the girls in the senior classes. Miss Molin was a kind, gentle person, but it was no fun to hear Mr Castegren bellowing at the "big ones". It was not a very big school and sounds carried. And then we knew that when we ourselves became "big ones" we were in for Mr Castegren.

In middle school I was taught by Miss Maria Svensson, a fine old lady whose family had been at Tumba Bruk for seven generations. Quite an idiosyncratic person, her special punishment being to make you put your hands on top of your desk, whereupon she would strike at the edge with her long ruler.

After intermediate school, it was time for us to be taught by Mr Castegren in upper school. He was an extraordinarily irascible teacher, and he made pets of children from the better-class families. He was not well-disposed to share-croppers' children from the surrounding farms or to us ordinary kids. Often he victimised us on account of our background. He disliked my father for being an active teetotaller, and so my brother and I were dubbed "sons of the Temple King", usually with a nasty smirk.

*The employees of Tumba Bruk at the beginning of this century. The Manager, Mr Fiebelkorn, is seated in the middle.*

During my childhood the mill was a patriarchal establishment in which the manager, Mr Fiebelkorn, held sway. Nobody, not even fellow-workers, was on christian-name terms. I know that several people working in the same team still did not call each other by christian names, and it was quite out of the question for workers to show any such familiarity to clerks. The guild system was still very strong, and it was terribly difficult to get anybody to teach you the various crafts - beating, forming and couching - as I was soon to find out when I left school and began working at the mill.

To begin with I worked in gardens belonging to private individuals or to the mill. Workers at Tumba had free accommodation, fuel and a garden plot, so for a small lad with an interest in gardening there was plenty of work, even though it was hard graft and badly paid. You received about two crowns a week for what was seldom less than ten hours a day, though Saturdays were a bit shorter. The job also included fetching firewood for the clerks. My mother helped the family to make ends meet by baking crisp bread which she sold to the employees at the mill.

Then in 1916 I became an apprentice at the papermill. To begin with I worked in the separating room, where the dried hand-made paper was separated from the interleaving paper. Everybody not employed on forming and beating was

*The finishing room, 1922.*

occupied there for a few hours each morning. There were quite a few of us there, and we had to hurry; this work mustn't take too long, because everybody had other tasks to attend to.

We started early in the morning by taking down the paper which had been hung up to dry in the drying loft the previous day. From the loft, which was next to the finishing room, the dried sheets were taken in for finishing and further treatment. When I had finished separating for the day, I often had to grate the edges of sheets which had been put in bundles of about 500. This was hard work and you had to be very accurate about it. The customers wanted sheets of equal sizes but still with the raw edges. This is why the paper could not be cut to the required format.

As an apprentice I also had to collect the mail bag to be sent by the 7.30 train to the Bank of Sweden in Stockholm. When leaving the bag I also collected the day's post for the mill, which I then had to go round distributing. At three-thirty it was time to collect a new mail bag from the station and distribute more post. Sometimes I did not finish the day until almost eight in the evening. I felt like a beast of burden.

Eventually I got out of the finishing room and started working as a layboy in the vat room. I wanted nothing more than to be able to advance in the trade and learn all the different skills - beating, forming and couching. The layboy was the least experienced member of the vat team, while the vatman had many years' training behind him. The vat team worked like this:

The vat is filled with whole-stuff diluted with water, after which the vatman

*Tumba Station, c. 1915.*

tests with his hand to see if the stock is the right concentration and temperature. It has to be between 30 and 40°C and free from lumps.

When the vatman is satisfied with the stock, he takes his mould and dips it into the stock, which at this stage of things resembles lukewarm gruel. The mould is a wooden frame strengthened with wooden strips laid crosswise. Two metal cloths are stretched over these strips. The lower cloth is sewn onto the strips, and the upper one is sewn onto the lower cloth. Any watermarks are sewn or soldered onto the upper cloth. Another frame, called the deckle, is placed over the mould, and this keeps the pulp from immediately running over the edges and back into the vat.

The vatman, then, dips the mould, with the deckle fitted to it, into the vat and takes up pulp, which is left lying on the metal cloth while the water slowly runs down through the cloth and back into the vat. If the vatman has taken up too much pulp on the mould, he dislodges the excess during his "stroke", a shaking movement which makes the fibres set better. It takes a great deal of skill on the vatman's part to shape one sheet after another, all with the same surface weight. When he is satisfied with the condition of the pulp, he removes the deckle and passes the mould to the coucher, who stands beside him, but holds onto the deckle and places it on the next mould, in which he begins making another sheet.

*The vat room, 1922.*

The coucher leaves the mould for a while in a tilting position, so that the water can drain off. He then presses the wet layer of pulp against a moistened felt on a slightly corrugated "couching board". When he lifts the mould again, the layer of pulp remains on the felt.

Mostly two moulds and one deckle are used for forming and couching. While the coucher is removing a sheet from one mould, a new one is being formed on the other, after which vatman and coucher exchange moulds and, during the few seconds which the water is allowed to drain away from the mould containing the new layer of pulp, the coucher "pitches" a new felt on top of the sheet which has just been couched. This felt will then receive the next sheet to be couched, and so it goes on until a post (150-200 sheets) has been completed.

The finished pile of felts and sheets - the post, as it is called - is then put into a hydraulic press, in which most of the water is squeezed out. Then it is the layboy's turn. He has to pull the newly pressed sheets away from the felts and pile the sheets on top of each other, very carefully. The sheets are then pressed once more, this time without any felts in between. The empty felts are returned to the coucher.

After the day's work is over, the sheets are taken to the drying room, where they are hung up on poles to dry.

69

Since my great ambition was to become a vatman, I worked quickly as a layboy, so that I would have a little time to spare before each pressing to learn the art of forming. I had prepared pulp in a tub and my father, who worked in the hollander room, had lent me a small test mould. As soon as I had a moment to spare I began practising the right motions - the "stroke" - with the mould.

My duties as a layboy also included straightening out felts, and being rather a maid of all work to the vatman and coucher. When the vatman felt that the pulp was getting too thin, he would shout for more pulp to be poured into the vat. The pulp was kept in one big stuff-chest, and it was fetched to the vats in a "shaft", a wooden bucket holding about 25 litres. When the bucket was not being used, it was left in a tub of water to keep it clean. Altogether a "shaft" of pulp weighed something approaching 50 kg, so you needed strong arms to lift it over the edge of the vat. After emptying the "shaft" I had to stir the vat with a rake or "hog" to stop the pulp from getting lumpy. You might think that the vatman could have taken care of the "shaft" and the pulp, but that was not the way in those days. The vatman was king and his word was law.

We layboys also had the job of waking the vatman so that he would not be late for work. This meant that we had to start an hour earlier. If you happened to oversleep or for some other reason arrive too late to wake the vatman, you were certainly given what for. A layboy at a papermill in Småland, in the south of Sweden, overslept one morning and was given a good hiding when the vatman arrived and found that the water in the vat was too cold. The layboy at that mill also had the job of making sure that the vat was warm. The boy ran off in a terrible state to his father, who worked in the hollander room, to tell him that he had been given a hiding. When the father was told the reason, he gave him one more.

New facilities were built in 1917 for the hollander, the finishing room and the vat room. My vat team was the last one to move into the new premises, which were opened in 1918. At the official opening ceremony, the manager, Mr Vestergren, who was quite a chemist, treated us to buns which he had made from wood pulp. He had ground the flour himself.

Hourly wages were introduced during the First World War. In 1920-21 a leading vatman had SEK 2:50 an hour and the second vatman and coucher SEK 2:30, but wages fell during the 1920s and 1930s, rising again during the early years of the Second World War.

The 1st April 1919 was a big day in the history of the paper-mill, at least for the workers, because that was when the eight-hour day came into force. An important reform which many people had not believed possible. The papermill workers' day seldom fell short of 12 hours, from Monday to Friday, and after work was over on Saturday afternoons they had to wash the felts and clean the

*Copper vat heater, "pistolet". Långasjönäs Bruk. Photograph: Göteborg Historical Museum.*

moulds, a job which formed part of their duties but could not be done during paid working hours.

Somebody wrote a special song commemorating the introduction of the eight-hour day at Tumba.

For my own part I had now begun working as a vatman and there are two expressions which I heard at the papermill and which have stuck in my mind. One of them is: "Well-boiled is half-bleached says the paper-maker", and the other is: "The hollander makes the paper". I shall be talking about the importance of the hollander for pulp quality later on, but first a few words about the washing and treatment of the raw rags.

There were three stages in the treatment of rags:
1.  Mechanical cleaning, sorting and maceration.
2.  Chemical cleaning or boiling under pressure.
3.  Production of half-stuff, which in the majority of cases meant bleaching as well as rough-beating.

The large rag-picking establishments, as well as the small rag merchants, delivered, it is true, specified rag qualities, but these seldom agreed with the paperman's requirements. Even the best rags in the market had to be picked over and re-sorted in more limited classes, in order for the factory to maintain its accustomed qualities.

71

The rags arrived in big bales of about one cubic metre. One of these bales could weigh anything between 300 and 400 kg. Some of them were grimier than others too, but all rags were dusty. Out of consideration for the workers' health, therefore, the rags were first put through a whipping machine to rid them of as much dust and sand as possible. The whipping operation fluffed up the rags, which until now had been compressed, and this made it easier to sort them afterwards. Between 2 and 9 per cent of the rag material disappeared during whipping, depending on how clean or dirty it was.

After whipping the rags were taken to the sorting room for final sorting. This work was usually done by women. In the sorting process, all buttons, hooks and other foreign items were removed. Impregnated parts, especially rubber, were also removed, because otherwise they could cause a great deal of damage in the manufacturing process. And pockets and seams were cut up, because they were dust traps. Then the rags were sorted according to fibre material, colour, degree of bleaching, wear and cleanness. Sorting was done at tables, called "hurdles", where most of the tabletop consisted of coarse iron wire netting. A ventilator under the netting extracted the dust.

Before the rags could be processed any further, they had to be turned into more or less equal pieces. This was done in a rag-cutting machine. Cutting, once again, generated large quantities of dust, and this was removed by putting rags through a dust extraction device.

After this the rags were boiled under pressure. In the old days the sorted rags used to be moistened and put into big piles. They would then start to ferment, and the rising temperature had the effect of detaching impurities and loosening the bundles of fibres. This procedure was superseded by boiling under pressure. Mechanical treatment of the rags only served to remove loosely attached impurities. To remove glue, the rags also had to be treated in water. Other contaminants such as oil, grease, dirt, resin, tar etc., on the other hand, could only be removed by alkali boiling.

During this boiling process the impurities were saponified, and in this form they could be rinsed away afterwards. Boiling also served the important purpose of removing, or at least reducing, the colouring of the rags, so that it could be completely obliterated by bleaching.

The importance of boiling can be seen from the story of how my father, at that time first man in the hollander room, solved the problem of the paper for the catalogue of the Hallwyl Museum. This tremendous order for a 78-volume work was one of the biggest the mill had received. Countess Hallwyl herself, no less, visited Tumba to discuss paper qualities. She was very particular about the paper being "flax-coloured". Presumably she had a romantic notion that it would resemble the colour of unbleached linen handkerchiefs.

*Sorting rags. Klippan Archives.*

My father decided to use rejected mail bags for the input rags, because they were made from flax with a jute weft. After many experiments with different boiling times and bleachings, he managed to produce a pulp and a paper which met with the Countess's approval. The fact of the flax colour coming from the jute in the pulp did not matter so much. This particular pulp was carefully bleached with a little chloride of lime, but it was very well boiled indeed.

Lastly, in the case of undyed, unbleached coarse rags, boiling also had served to dislodge encrusted substances holding the fibres together.

Soda, sodium hydroxide and lime were used as agents in boiling. The agent to be used was decided ad hoc. Lime was administered in the form of lime-wash and, although diluted, was capable of destroying encrusted substances, though without dissolving them completely. On the other hand, lime was excellent for destroying many dyes. It could seldom be shown to have weakened the fibre material. This would happen, for example, in cases where large quantities of lime had been used in the prolonged or intensive boiling of clean, white rags. In this case the lime would not have found any substances to saponify with, and it would therefore act all the more energetically on the fibre cellulose.

Lime boiling was mostly used for coarse and, especially, heavily dyed rags. On the other hand it often proved insufficient for boiling rags containing tar and oil or linen containing a lot of shives.

# A device for cleaning unsorted rags.

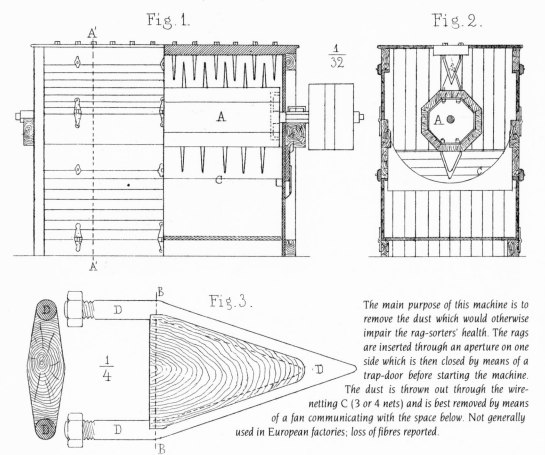

## Fig. 1.

$\frac{1}{32}$

## Fig. 2.

## Fig. 3.

$\frac{1}{4}$

The main purpose of this machine is to remove the dust which would otherwise impair the rag-sorters' health. The rags are inserted through an aperture on one side which is then closed by means of a trap-door before starting the machine. The dust is thrown out through the wire-netting C (3 or 4 nets) and is best removed by means of a fan communicating with the space below. Not generally used in European factories; loss of fibres reported.

## Pre-sorting list for rag purchasing.

| | K. g. | pris | Kronor | öre | | K. g. | pris | Kronor | öre. |
|---|---|---|---|---|---|---|---|---|---|
| Bleached linen | | | | | Cordage netting | | | | |
| Bleached cotton | | | | | Tarred rags | | | | |
| Half-bleached linen | | | | | Flax-spinning waste | | | | |
| Unbleached linen | | | | | Cotton waste | | | | |
| Dyed linen | | | | | Oakum | | | | |
| Dyed cotton | | | | | Tow | | | | |
| Dirty rags | | | | | Knitting wool | | | | |
| Canvas no. 1 | | | | | Wool half-wool | | | | |
| no. 2 | | | | | Paper chips | | | | |

# Rag-cutting machine.

## Fig. 1.

$\frac{1}{24}$

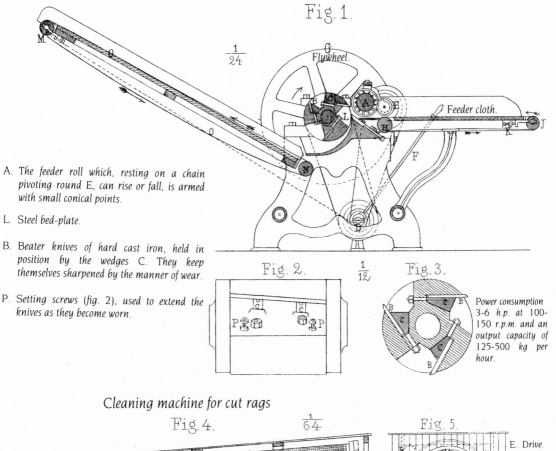

Flywheel.

Feeder cloth.

A. The feeder roll which, resting on a chain pivoting round E, can rise or fall, is armed with small conical points.

L. Steel bed-plate.

B. Beater knives of hard cast iron, held in position by the wedges C. They keep themselves sharpened by the manner of wear.

P. Setting screws (fig. 2), used to extend the knives as they become worn.

## Fig. 2.

$\frac{1}{12}$

## Fig. 3.

Power consumption 3-6 h.p. at 100-150 r.p.m. and an output capacity of 125-500 kg per hour.

## Cleaning machine for cut rags

## Fig. 4.

$\frac{1}{64}$

## Fig. 5.

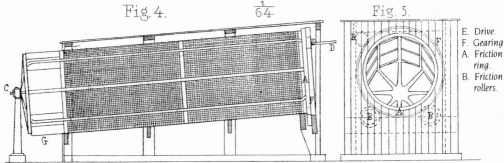

E. Drive.
F. Gearing.
A. Friction ring.
B. Friction rollers.

Consists of a slightly tilted, octagonal sieve, which when rotated shakes the rags between its walls (iron gauze no. 3-4), so that the rags are gradually pulled down, by the force of gravity, from its top to its lower end.

The housing surrounding the prism should be connected to a fan. The speed of rotation adapted so that the rags are thrown with the utmost force between the walls of the prism. The rags are inserted at A and drop out at G. Loss during cutting and cleaning 6-9% for fine and 10-15% for coarse rags. The rag-cutting machine and rag cleaner should be positioned above the boiling room, so that the rags can be shovelled straight down into the boiler. The rag-cutting machine, which operates by vibration, should be placed on a firm base.

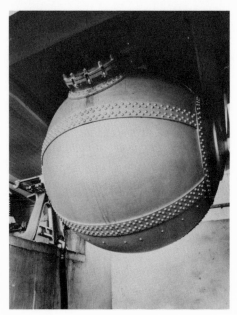

*Rag boiler. Klippan Archives.*

Soda boiling was reserved for finer rags which were white and less contaminated. True, soda can also be used with more contaminated rags, but this is uneconomical because of the excessively large quantities which are then needed.

Sodium hydroxide boiling, finally, was used for oily, tarry and dressed rags. Sodium hydroxide boils more vigorously than lime and ordinary soda and dissolves encrusted substances and also oil, grease, proteins (from dressing) and quite a few dyestuffs.

The general rule is not to use more alkali in boiling than will result in a clean and bleachable material. The amount of lime added could vary between 3 and 15 per cent, and soda and sodium hydroxide inputs between 1 and 5 per cent. The results of boiling depended not only on the concentration of the lye but also on pressure and boiling time. Since pressure and concentration made more difference than boiling time and since the fibre was easily weakened by excessive pressure, boiling tended to take place at low pressures (1 kg - 2 kg) and with long boiling times (3-12 hours).

At first boiling was done in open vessels over an open fire, but before long enclosed boilers were constructed which operated at a pressure of about 2

*The pictures on the previous two pages come from an anonymous MS in the author's possession, probably dating from the early years of the 20th century.*

*Emptying the rag boiler. Tumba Archives.*

atmospheres. Direct firing was eventually superseded by steam heating. The rotary spherical boiler which remained in service virtually unaltered until the 1960s was first introduced at the end of the 19th century. It was originally intended for boiling straw, but gradually it came to be used exclusively for boiling rags under pressure, because this way the rags mingled better with the lye. The boiler rotated at a speed of -1 r.p.m. The steam was introduced through the shaft pivots on which the sphere rested.

After treatment in the boiler, the rags were supposed to be soft and pliable and somewhat discoloured. The entire charge - all the rags in the boiler - had to be equally discoloured. If they were not, this would mean that the rags and lye had not been sufficiently mixed.

If there were shives in the rags, these had to be sufficiently soft and enclosed; the paper-maker would test for this by tearing a few specimens. If a boiling had turned out badly, e.g. due to oil and tar dyes not being sufficiently dissolved, or because the rags were still hard, the results could be improved by putting the rags into a wet pile for several days, during which time they would ferment.

However carefully one sorted the rags, certain fibres were liable to go undetected. Often these fibres changed colour during boiling. Jute, for example, turns dark yellow or brown. If this happened, the rags might have to be re-sorted after boiling. This "wet sorting" was also a way of checking up on the sorting which had been done previously.

# Boilers.

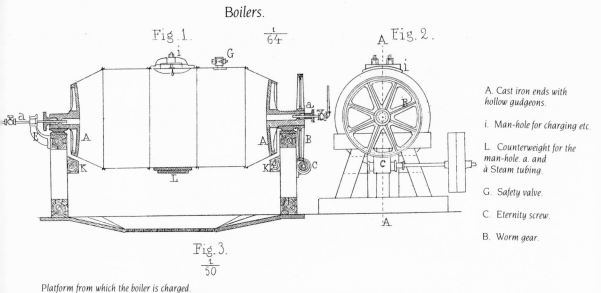

Fig. 1.    $\frac{1}{64}$    Fig. 2.

A. Cast iron ends with
   hollow gudgeons.

i.  Man-hole for charging etc.

L.  Counterweight for the
    man-hole. a. and
    à Steam tubing.

G.  Safety valve.

C.  Eternity screw.

B.  Worm gear.

Fig. 3.
$\frac{1}{50}$

Platform from which the boiler is charged.

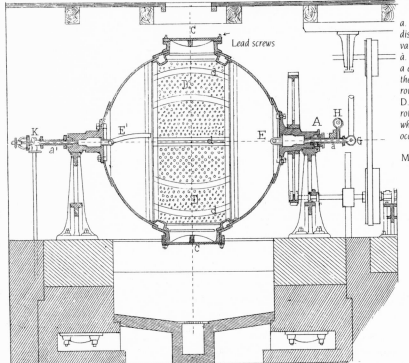

Lead screws

a. Induction tube, which can be connected to or
disconnected from the steam tubing by means of
valve H and to/from the water supply by valve G.
à. Drainpipe with stop valve K.
a and à are connected with the screening drum by
the diametrically opposed tubes E and E' bent
round the wall of the boiler.
D. When steam or air is to be led off, the boiler is
rotated so that E' occupies its uppermost position;
when water is to be squeezed out, it is rotated till E'
occupies its lowest position.

Many consider the screening drum superfluous.

Boilers are often fitted on the inside with rivetted angle brackets which induce the rags to participate in the rotation. Boilers
should be fitted with a safety valve and lead valve. Boilers of 4.2-9.2 cubic metres capacity will hold 500-1,500 kg rags.
Boiling time 2-15 hours. Rotation speed 1-2 r.p.m. with a power consumption of 1-2 h.p. The pressure in rag boilers does
not exceed 4 atmospheres; in boilers for straw or chemical pulp it is often 11 atmospheres.

When the boiler was emptied, some of the dissolved impurities accompanied the boiling lye, but most of them still clung to the boiled rags and could only be removed by thorough washing.

If the boiling had been a success, there was now every chance of producing first-rate rag pulp. But a lot of work remained to be done before the boiled rags had been converted into pulp. First, not only must the fabric be dissolved into its elementary constituency and the latter into individual fibres, but the individual fibres themselves had to be shortened, made pliable and, to a greater or lesser extent, split down their length.

Rags were originally macerated by pounding them by hand in mortars. Later on stamping mills were invented. These were powered by water or wind and they consisted of a long trough in which the rags were beaten with hammers or stampers.

Then, in about 1670, a machine was invented which, in one blow, made the stamping mill superfluous. The invention was made in Holland, or at least that is where it was first introduced, and so the machine is known as a "hollander".

Before going into any further detail about the way in which the hollander makes the paper, I will tell you a little about my life as a vatman at Tumba Bruk and conditions there. The first rule about holidays for workers at the mill was included in the 1919-1920 wage list. Workers of up to three years' standing were given six days' holiday, and those of longer standing were given 10 days. Because I was so interested in things technical, I exchanged my bicycle - it had drop handlebars - for a Husqvarna motor-bike. A year or so later I swapped again, this time for a new D.S. motor-bike.

Then, after completing my military service by stages between 1922 and 1925, I at last had the chance of taking my life's first holiday. I went off to Dalarna with a girlfriend. Petrol pumps were few and far between in those days and the gravel roads were not up to present-day standards, but my D.S. went like a dream. The Museum of Technology in Stockholm has a specimen of this motor-bike, which did not remain in production all that many years. D.S. was short for "Dark Special", and the bike had been designed by an engineer called Dark and a vet called David Sening.

By 1925 I was a vatman at Tumba Bruk and was greatly trusted by the management in spite of my relatively tender years. Johannes Vestergren was in charge at Tumba nearly all the time I was there, and he backed me up a great deal when the older vatmen wanted to stick to the old guild practices.

The guild mentality meant that you had to switch from a "senior", better-paid job to a more junior one if an older worker came along who had precedence. There is one episode I remember from that time. A vatman came home after having been away in America for some time. He had previously worked at the papermill as a vatman, so they could not make a layboy of him. Being the youngest, I had to give up my position and go back to working as a layboy. But it was not many days, perhaps a week, before the newcomer developed a swelling of the joints. He could no longer continue as vatman, and so I got my previous job back.

I liked my work as a vatman and I liked working at Tumba. Then again, I was employed by the Swedish Government, so my security was assured.

I got to know my Hulda in 1925 and we were married in 1927. We got a flat - one room and a kitchen - in the old smithy, on the site of what is now the sewage processing plant.

In the mid-1920s I was earning SEK 1:55/hour, whereas in 1920-21 it had been SEK 2:03. Wages fell, in other words, and they did not level out until the end of the 1930s. The world depression made itself felt at Tumba, and between the autumn of 1930 and the summer of 1932 we were put on a five-day week.

The question of retirement for the oldest workers at the mill cropped up at the same time, and in 1932 the Riksdag (parliament) passed a law on the subject. Nine old stalwarts were given pensions that year. One of them, P.J.O. Tyrenius, had been at the mill for 76 years.

During one of the rounds of pay talks in the 1930s, Per Edvin Sköld, at that time Inspector of the papermill, said that he would have to reduce our hourly rate by 5 öre in order to save jobs for the youngsters. We then counted as a high-paid group, because we were making SEK 1:55 an hour. We agreed to his demands, so that the youngsters could keep their jobs, but the only result was that we found ourselves making 5 öre less per hour. The youngsters were laid off anyway.

Although we had our ups and downs at the papermill, I was always glad to go off to work together with my mates in the vat team. As I told you earlier, most vat teams did not have much in the way of team spirit. The old vatmen clung to their guild privileges and would not deign to talk to their younger colleagues, still less to call them by their Christian names.

In our team, things were quite different. The first vatman, Wenzel Andersson, was a kindly soul who realised that even a younger colleague was capable of doing good work. Thanks to his frequent encouragement, we young ones believed in the craft, but it earned him quite a few sour looks from the other gaffers.

Hulda and I thought we had had enough bumping around on the motor-bike in 1928, and so we swapped it for a model T-Ford. That stayed with us until 1931, when we invested our savings in a brand new model A-Ford, costing SEK 3,400. Our daughter Anna-Carin was born on New Year's Eve 1931, and we felt more firmly rooted in Tumba than ever.

Work at the mill ran smoothly until the beginning of 1934. My father, Anders,

*D.S. and rider.*

at that time first man in the hollander room, was due for retirement and I was asked if I would like to take over his job. I had had no practical experience of working in the hollander room, but Dad and I often talked about our different jobs, so I had a fair idea of what it was all about. Even so, I was not all that keen on the job. I was so happy with my mates in the vat team and I was unwilling to leave them for a completely new job that I didn't know that much about. And there was still the old divide between young workers and older ones.

But in the end I gave in and promised to succeed my father in the hollander room. He was to stay on for a month after his retirement to help me get the hang of things.

It was part of the first man's job to requisition raw material from the store and to order boiled material every day, to the engineers' recipes. It was very important to keep a list of coarse and fine-ground pulp, so that you could obtain the right sorts for all the production.

The boiled material was taken to a hollander to have the impurities washed out. The commonest type of hollander is an almost oval trough. Along the middle is a partition wall, which does not reach all the way to the sides of the trough but leaves a gap on each side so that the contents of the Hollander can circulate round it. One side of this dividing wall, right across the trough, is a roll fitted with knives. Underneath this roll there is a bump in the bottom of the hollander - the bedplate - which is also fitted with knives.

81

82    *Two pictures from a typical worker's home at Tumba Bruk in 1922. Tumba Bruk Archives.*

When the roll, which can be raised and lowered, begins to rotate, the pulp is set in motion and, by regulating the distance between the roll and the bedplate, one can grind the fibre as coarsely or finely as required.

But before the material could be beaten, it had to be washed. This was done into washing beaters, fitted with a washing cylinder. Before beating the rags were made to circulate for a certain length of time under a continuous stream of pure water and escape of dirty water through the washing cylinder. The latter, which was lowered into the macerated rags during washing, consisted of a drum fitted with fine metal gauze. When the waste water began to look clear, the roll was lowered closer to the bedplate and half-stuff beating could begin but with washing still continuing.

The purpose of half-stuff beating, then, was to turn the boiled and perhaps washed rags into a pulp of detached fibres which would be easy for the bleaching liquid to get at. After half-stuff beating, the fibrous pulp could very well be subjected to the special final beating for the production of a certain kind of paper. Final beating was done in the whole-stuff hollander, but even in the half-stuff hollander the rags should be treated according to the type of paper which is to be produced.

Good half-stuff must be of absolutely even consistency, which means that it must not contain any unmacerated bits of rag or any large quantity of short fibre fragments. The beater man supervised the beating by lifting up some of the pulp on his testing rod and checking to see how far the disintegration of the rags had progressed.

Despite previous sorting and cleaning, the rags could still contain gravel, litter, buttons and suchlike when they reached the hollander. To trap such foreign bodies, the hollander was fitted with a sand-trap immediately upstream of the roll. A sand-trap of this kind consisted of a declivity in the bottom of the hollander, covered by a perforated sheet of metal which was countersunk, so as not to rise above the bottom of the tub. Large particles or objects, such as nails, nuts and so on, which could damage the knives if they got between the roller and the bedplate, were trapped in a channel in the bottom of the hollander, downstream of the sand-trap.

Washing began immediately after the rags had been loaded into the hollander. This started with the roll raised, so that the rags were merely moved round the trough through the washing drum, without coming into contact with the roll and bedplate. It was absolutely vital for all the dirt to be removed, so that impurities - from mail bags, for example - would not be ground in. The washing drum rested on journals arranged in such a way that the drum could be lowered to different depths in the contents of the hollander but could also be lifted clear of them.

When the drum was lowered into the contents of the beater, the dirty water

*Laystool. Photograph: Göteborg Historical Museum.*

passed through a screen or "chess" stretched round the drum and was then picked up by paddle boards inside the drum and lifted towards the centre of the drum, escaping through the outer side wall of the drum. To prevent the contents of the beater getting too dry, fresh water had to be constantly applied behind the drum.

When no impurities could be detected in the escaping water any more, the real beating process began. Different types of fibre had to be beaten for different lengths of time. The ordinary beating time for old cotton, for example, was about three hours, while linen rags, which were much harder, could take anything between five and six hours. Thus the washing and beating work which had started at crack of dawn first began to show results in the afternoon.

When the beater man judged the pulp to be ready, it was discharged into big chests - one for each grade of coarse pulp. These chests were built of brick, with scupper holes for the excess water to escape through.

Boiling and washing had now removed the impurities which could be detached or destroyed by alkali, but there still remained dyestuffs, either those typical of the fibre in its original state or those which had been artificially added in a dyeing process. To produce a completely white and pure material which can be used for white, pale-coloured paper, the half-stuff has to be bleached, even if it has been made from white rags.

*Hollander with tiled trough.*

From he big chest the half-stuff had to be transferred to a bleacher. What happened was that a bleacher boy - the third man in the beater room - raked up the pulp into wooden boxes holding about 450 litres. These were then moved on trolleys to the beater, which could take about two cubic metres.

The rag fibres were bleached with chlorine water, which was stored in a tank and piped to the bleacher. It took about 90-100 litres of chlorine water to bleach a batch of old cotton pulp, while for the same amount of linen rag pulp you needed three times as much. This kind of bleaching occupied a whole working day and was done by the bleacher boy single-handed.

The bleacher was a special kind of beater with neither roll nor knives, used for bleaching and nothing else. Instead of the roll forcing the pulp round in the tub, paddle wheels or worm-gear wheels were used. The contents of the bleacher were heated to speed things up, and also to improve the whiteness of the product. The pulp temperature would be about 30-40 degrees.

From the bleacher the bleached half-stuff was transferred to chests resembling those used for the coarse stock.

The bleached pulp was not considered ready until it was entirely free of chlorine residues. To check this, it was tested with iodine/potassium solution. A small quantity of pulp was extracted and a few drops of the solution added to it. If it turned blue, the pulp was not entirely free from chlorine, and antichlor then had to be added to remove the residue.

# Half-stuff beater holding 200 kg (13 h.p.)

## Fig. 1.

$\frac{1}{40}$

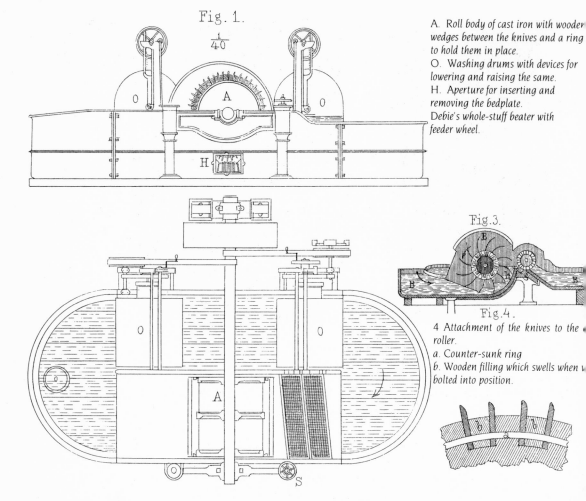

A. Roll body of cast iron with wooden wedges between the knives and a ring to hold them in place.
O. Washing drums with devices for lowering and raising the same.
H. Aperture for inserting and removing the bedplate.
Debie's whole-stuff beater with feeder wheel.

Fig.3.

Fig.4.

4 Attachment of the knives to the roller.
a. Counter-sunk ring
b. Wooden filling which swells when bolted into position.

## General particulars concerning beaters

Beater tub of wood, cast iron or cemented masonry. Charge 400 kg rags for 5.5 x 3 m. half-stuff tub, 50-60 kg for 3.2 x 1.6 m. half- or whole-stuff tub. Roll body of cast iron or else an oak block, turned and wedged into position (fig. 4).

Roll diameter 0.527-1 m, no. knives 38-56 in half-stuff (distance between knives here not less than 60 mm) and 48-72 in whole-stuff chest, speed 160-200 revolutions (6-9 m/sec peripheral speed); average power requirement 10-15 h.p. for half-stuff, 6-8 for whole-stuff rolls, weight of roll 500-1,000 kg or more.

Roll knives 10-15 mm thick in half-stuff, 8-10 in whole-stuff rolls.

The roll is freely journalled (without any hood) in its bearings and acts by its own weight only. In more recent beaters the two gudgeons can often be raised or lowered. The process of beating the rags into ordinary half-stuff takes 2-3 hours, and it takes 2-3 hours to beat the half-stuff into ordinary whole-stuff. A much longer beating time is needed (i.e. the roll has to be lowered less rapidly) when strong, fine paper is desired.

# The beater tub with washer discs

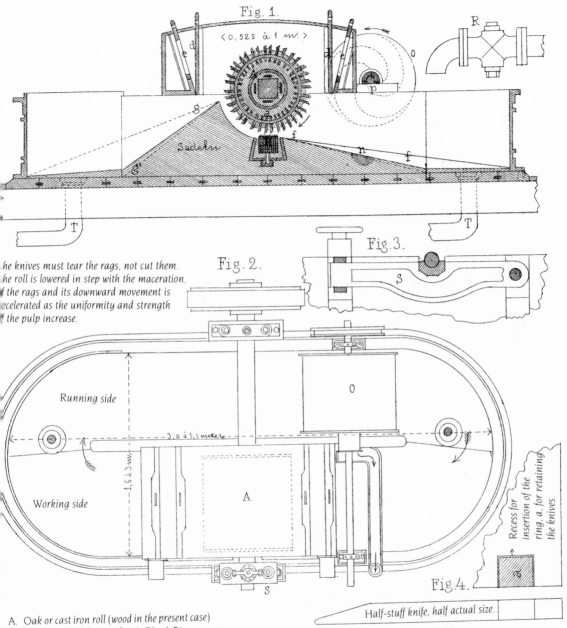

**Fig. 1.**

< 0,525 à 1 m. >

**Fig. 2.**

**Fig. 3.**

**Fig. 4.**

Running side

Working side

3,2 à 5,5 meter

1,6 à 3 m.

Recess for insertion of the ring, a, for retaining the knives.

Half-stuff knife, half actual size.

he knives must tear the rags, not cut them.
he roll is lowered in step with the maceration.
f the rags and its downward movement is
ccelerated as the uniformity and strength
f the pulp increase.

A. Oak or cast iron roll (wood in the present case) with steel or metal knives (fig. 4, PL. 4-5).

H. The bedplate, with 10-20 knives set closely together.

e. The wash discs, fitted with metal gauze which will allow the dirty water to escape but will contain the rags; recently often replaced with wash drums.

d. Washer doors, lowered when washing (circulation of water) is to be concluded.

O. Cylinder washer (see also PL. 5-6) with metal gauze casing and spiral paddles. The former separates the dirty water from the rags, the latter transfer it to the centre and the drainage channel. The washing cylinder is lowered or raised (see Pl. 5) depending on whether it is to take up a larger or smaller quantity of water or none at all.

R. Cock for admission of pure water.

S. Hoisting gear for raising or lowering the roll; see also fig. 3.

T. Drain pipe with valves.

*Beater and washing cylinder*

During the thirties at Tumba we had three breaker beaters, four bleachers and six whole-stuff beaters. To keep the whole-stuff beaters supplied with ready-bleached, washed pulp, the bleacher boys collected chests of pulp according to the engineers' recipes. The whole-stuff beater took four chests, each with a dry weight of 12 kg.

If one day I would be beating pulp for watercolour paper, then I would order the following from the bleacher:

3/4 chest no. 3 (old, semi-bleached linen) = 20%
2 1/4 chest no. 9g (old bleached cotton) = 55%
1 chest no. 9 new (new bleached cotton) = 25%

and then 20 litres of resin size and 2 kg of alum would also be added to the charge.

Beating this pulp took three hours.

Now it is above all in the beating of whole-stuff that the paper-maker can demonstrate his skill, no matter how necessary it is to master the other operations in order to produce good pulp economically. Although the properties of the paper are supremely dependent on the whole-stuff beating, it is remarkable how little is known about the workings of the whole-stuff beater. This is very much because there are so many factors involved in the beating that some of them were overlooked and the relationship between them and the beating result cannot always be expressed or determined exactly.

The fibres in half-stuff are still for the most part quite long and sometimes even matted into threads. Any attempt to produce paper straight away from this pulp, without any further beating - straight out of the stuff chest, in other words - would not result in anything resembling good paper as regards evenness, felting or other properties.

The half-stuff has to be further macerated and dissolved. The first task of the half-stuff is to completely isolate each elementary fibre.

Since the skill of the paper-maker is very much a question of utilising the capacity of the fibres for felting together, his work would be made more difficult if not impossible were each fibre unit to be completely isolated in the pulp. The fibres, which may be rough and rigid to begin with, have to be made more pliable and soft, so that when deposited on the wire they will be matted together in a more or less dense layer of pulp. This is an important task for whole-stuff beating.

Depending on the paper to be produced, the fibres also have to be shortened to some extent. All these processes - separation of the fibres, softening and shortening - take place at once. But beater processing can be varied so as to make one or another of these results of beating predominant over the others.

The knives in the beater, which are fitted on the roll and in the bedplate, bring about changes in the fibrous material, e.g. by flattening and crushing. Often this results in a dissolution of the fibrous elements, causing them to disintegrate longitudinally into smaller units, known as fibrils. If beating is driven to extremes in this latter respect, as is often the case, for example, in the production of first-class rag paper, bank note paper or cigarette paper, the fibres are to a great extent beaten out of all recognition.

In the beating process one can basically tell whether the fibres have been cut, chopped or shortened between the working knife edges or whether they have been crushed, compressed or torn off between them. The character of the ready-beaten pulp will depend supremely on the most prominent characteristic of the beating process - cutting, crushing or tearing. And it is impossible to describe the beating capacity of a beater without specifying the character of the ready-beaten pulp.

The paper-maker has two principal terms for the character of the pulp: free stock and greasy stock. If the beating has mostly involved cutting and shortening the fibres, then you say that the pulp is free or quickly beaten (rösch in German, maigre in French). The characteristic of free stock is that, when it is transferred to the wire cloth, the water drains off easily.

The opposite of free pulp, then, is greasy pulp (schmierig in German, grasse in French). Greasy pulp has been differently characterised by different writers. Thus reference is made to a pulp obtained by crushing, fibrillation, water

# Wooden washing cylinder.

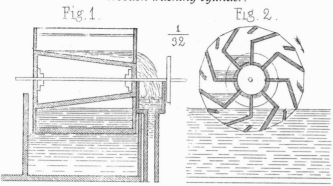

### Fig. 1.

### Fig. 2.

$\frac{1}{32}$

### Fig. 1 och 2.

Washing cylinders operate less vigorously but cause less fibre wastage than washing discs and have therefore come to be widely used recently. They should not be interposed between the water pipe and the roller, because then they lead off the pure water.

# Kingland's centrifugal beater

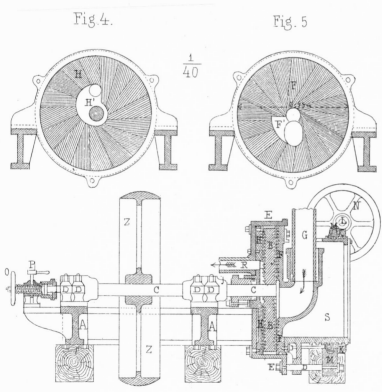

### Fig. 4.

### Fig. 5

$\frac{1}{40}$

### Fig 4, 5 och 6.

E. Cast iron body, against the fluted, chi[...] bedplates F and H of which a fluted, chilled [...] iron disc B acts while rotating at high speed.

G. Tube through which the slurry-like pulp [...] hydrostatically forced into the mill, passing rou[...] B and then escaping through R.

O. Shifter handle for adjusting B in relation [...] H.

N. Shifter handle for adjusting F in relation to [...] The centrifugal beater, contrary to what w[...] originally hoped, has not proved capable [...] replacing the whole-stuff beaters. On the ot[...] hand it is used a great deal for mixing and [...] beating of whole-stuff. A beater of this kind can [...] the work of 6-8 whole-stuff beaters and requi[...] 10-30 h.p., depending on the fineness of the p[...] and the beating pressure.

*Wooden half-stuff chest. Holmens Bruk, Motala. Photograph: Göteborg Historical Museum.*

absorption, and hydrated cellulose formation. The beater man judges the greasiness of the pulp by touching it.

Greasy pulp has a property which is supremely interesting to the paper-maker. It retains water, in the sense of being slow to release it when the vatman has transferred pulp to the mould. This enables the vatman to shake the mould thoroughly, before so much water has drained off that the pulp stops felting.

Generally speaking, ideas concerning the working of the beater have been rather vague. True, an experienced paper-maker can tell whether the pulp in a beater is ready or not by looking at it and feeling it. But the same person may have great difficulty in describing at all closely the physical changes which it has undergone.

Often the beating process is explained by saying that, if the contents of the beater have a thin consistency and the roll pressure is heavy, the pulp will mostly be free. With a thick consistency and less roll pressure, the fibres are treated more gently and are only partly shortened and crushed (greasy pulp). This latter method results in the splitting of flax and cotton fibres, which is the same thing as fibrillation.

Careful investigation will show, however, that this explanation does not always hold good. If, for example, one puts a piece of spinning paper under the

91

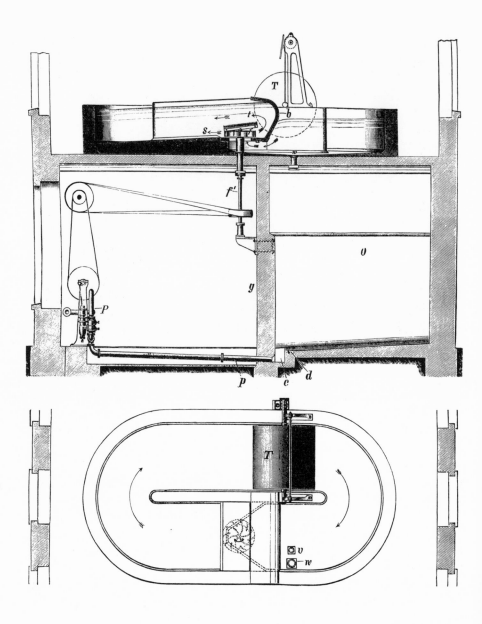

*Bleacher. The turbine impels the pulp towards S. T is the washing cylinder, driven by the movement of the pulp. Length 7 m, width 3.5 m. Capacity 800-1,000 kg pulp. From Hofmann: Papierfabrikation.*

92

*Oak beater roll with steel blades. Wookey Hole Paper Mill.*

microscope, the fibres in the sheet will appear to be virtually undamaged. It is almost impossible to detect any difference between the fibres in the sheet and those in the raw material. And yet the sheet must be produced from a very greasy pulp in order to achieve its particular character. The above quoted explanation of the beating process also fails to show why a sheet made from greasy pulp is thinner and more compact than one made from free pulp. Nor does it tell us why the pulp gets thinner when its temperature is raised. When the paper-maker has the pulp on the mould, the water drains off at different speeds, depending on whether the pulp is free or greasy, but this is no logical consequence of the explanation quoted above. That explanation, then, only conveys part of the truth and is, on the whole, unsatisfactory.

In other words, it is hard to give an exact forecast of the outcome of the beating operation before it has started. The skill of the beater man makes all the difference.

93

When the whole-stuff engine begins working, the primary result is to separate the matted fibres of the pulp which has previously been put through the breaker engine. At the same time, water mixes with the material and is absorbed by, or penetrates, each individual fibre. The fibres are cut and torn off when they pass between the roll and the knife of the bedplate. The thinner the layer of pulp passing over the knives and the higher the pressure, the heavier the cutting will be. One might imagine that no fibres would be cut off without passing the roll at such heavy pressure that the roll and bedplate were practically in contact with each other, but this is not the case. Even if the roll "floats" on a layer of pulp instead of butting against the bedplate, the fibres are still clipped to a certain extent.

The fibres, then, absorb more water when they come under pressure and are torn between the roll and the bedplate. The cell wall absorbs water, all the bores of the fibres are filled with water and the cellulose is converted into hydrocellulose. While this is happening the cellulose swells. Water uptake also makes the pulp slow to release the water later on.

The pressure and fragmentation to which the pulp is subjected during beating result in fibrillation and disintegration. Certain fibres are completely broken down into fibrils, while others are just torn up at the ends and others again are merely torn apart. As a result, the pulp will also include a number of small fibrous fragments, so that when the pulp is poured out on the wire, the interstices between the fibres will be filled by small particles and fibrils. This will make the layer of pulp denser and, for this reason too, the water will run off less rapidly than with less thoroughly beaten pulp.

A skilled beater man had to be able to control the beating so that the fibres would be shortened without making the pulp greasy. The process could also be controlled so as to make the pulp supremely greasy, without shortening the fibres. Beating had to be able to give the pulp any beating character whatsoever between these two extremes.

The way in which beating had to be conducted depended on the properties required in the finished paper. The main decisive properties were: strength; density (specific gravity, permeability to gases and liquids); look-through; and surface quality.

Depending on the fibre material, one property or another could be elevated by beating, but mostly at the expense of other properties.

Now of course, the quality of the pulp is dependent not only on beating but also on the type of fibre and the state of the fibrous material when beating begins. Different kinds of fibre, for example, have very different tendencies to longitudinal disintegration. Ramie (China grass), hemp and flax are among the easiest to fibrillate. Jute, manilla, wood cellulose and cotton are a lot more

*Hollander beater. Holmens Bruk c. 1890. Photograph: Göteborg Historical Museum.*

difficult and esparto and straw cellulose are almost impossible to split lengthwise.

This classification applies to well-preserved, unlignified or only slightly lignified fibres. If, on the other hand, the fibres have been weakened by chemical influence (alkali, chlorine etc.), this reduces the prospects of splitting them lengthwise to obtain a greasy pulp.

The factors determining the work of the beater were as follows:

the pressure to which the fibres were subjected between the knives of the roll and those of the bedplate,
the thickness of the pulp,
the movement of the pulp or the speed at which it circulated in the beater,
the peripheral speed of the roll,
the width and sharpness of the knives,
the material the knives were made of,
the beating time etc.

The general rules of beating were as follows:

maximum roll pressure, sharp knives and thin pulp in the roll (low pulp concentration) gave a free pulp,
slight and very gradually rising pressure gave a thick pulp,
and wide knives produced a greasy pulp.

95

With a certain kind of beating, the shortening of the fibres will depend on the beating time, but of course, fibres are more rapidly shortened under heavy roll pressure.

A skilful beater man had to be able to control the pulp so as to achieve the requisite greasiness at the same time as the fibres had been reduced to the requisite shortness.

I hope now that this account has made clear the meaning of the adage "the hollander makes the paper".

Sizing also varied according to the kind of paper to be produced. The purpose of sizing is to make the paper impermeable to liquid. In the thirties, and earlier, when people mostly wrote in ink, unsized paper would have been impossible to use because the ink would have immediately soaked into the paper in blobs - something like the way it does in blotting paper. Paper was also sized in cases where it was not absolutely necessary, so as to improve its firmness, hardness and rattle and to make it look nicer.

Sizing was performed before or after the conversion of the pulp into paper. In the former case, the size was added to the whole-stuff. This procedure was called stock sizing or, because it involved the use of vegetable sizes, rosin (resin) sizing. Nowadays, with most pulp systems having no beaters, the size is added, for example, through a mixing chest or box or a mixing pump.

In the latter instance the sheets were dipped into a solution of animal glue (made from hides or bones). The sizing solution was then squeezed out and the paper re-dried. This procedure was called animal or surface sizing. Surface sizing can give a very equal-sided paper with good properties.

Several factors help to decide the amount of size to be added. They included the type of pulp, the use to be made of the paper, the sizing material employed, the volume to be beaten, the temperature of the pulp suspension, the drying process, calendering etc. People spoke, for example, in terms of fully sized, half-sized and quarter-sized paper.

. To dissolve brown rosin size, 9 kg of soda were put in a dissolving chest with 20 cm of water and after this, 15 kg rosin was slowly added, under constant agitation. The batch was then kept on the boil until thoroughly dissolved. This took 45 minutes to one hour, after which the batch was transferred to two tubs and left to stand until the size had separated from the lye. The lye was then poured off and the rosin solution returned to the dissolving chest, where it was brought to the boil again before being strained into a tank. The entire batch occupied 46 cm = 530 litres.

It is interesting to note that this particular stage of paper-making, sizing, was the paper-maker's biggest headache in the old days. The Dutch masters employed at Tumba in the early years had promised to teach native Swedes all

*Vat with mixing stick. From the old Fröåsa mill (end 19th century). Photograph: Göteborg Historical Museum.*

the skills of the trade, but it was found that they had been able to keep their knowledge of sizing to themselves, even though they had been working at the mill for 20 years. In the end the Bank Commissioners found themselves obliged to promise the Dutchmen special privileges for themselves and their relatives, in return for revealing the secrets of sizing.

At the same time as I changed to the beater room in 1934, the Government Auditors noted that hand-made bank note paper in 1933 had cost 28-30 crowns per kg, whereas in other countries it was far cheaper. Denmark and Finland were the only European countries apart from Sweden which were now using hand-made paper for their bank notes. To save expense, the auditors felt that bank note paper would have to be made by machine instead. The Bank Commissioners had already mounted an inquiry, and they had appointed a special committee which, among other things, had been in touch with the German bank note paper mill at Spechthausen and the Escher Wyss machine factory in Ravensburg. In addition, the manager of the Tumba Papermill, Mr Wallentin, had been doing the rounds of Europe, studying different types of machinery.

97

The Bank Commissioners were convinced by foreign experience that bank note paper could be manufactured by machine and with no loss of security, and after a full inquiry they resolved, in April 1938, to install a paper-making machine together with the necessary equipment and to enlarge the factory premises, at an estimated cost of SEK 486,000.

Instead of enlarging the main factory building, as had been planned in 1916, a machine room was put up next to the forming room of the main building, on the south side. The result was a party wall between the two facilities. Building operations proceeded during 1938 and 1939.

In the summer of 1939, when work on installing the paper-making machine was in full swing, Hulda, Anna-Carin and I set off on a motoring holiday organised by the Union of Temperance Drivers of Sweden. This was a real baptism of fire for our model A-Ford, which had now been going for seven years. Our tour was to begin in Sundsvall, and on the way there we were going to stop off at Furuvik and have a look at the zoo. That part of the programme was literally a wash-out; it rained so hard that we couldn't leave the car. But the rest of the journey was an experience to remember. From Sundsvall we travelled north by way of Umeå, Piteå, Luleå and into Finland, continuing from there to Narvik and then down through Norway home.

We had just about got home again when the Second World War was upon us and the postman brought me a mobilisation order. I was called up on 2nd September, but later the order was countermanded and the Tumba personnel were exempted from military service.

There were several German engineers and riggers at Tumba while the paper-making machine was being installed, and as soon as it had been successfully commissioned and the first paper delivered, they were ordered to report for military service.

Production of bank note paper soon got started, although not full-scale to begin with. The first bank notes printed on machine-made paper were the ten-crown notes issued in 1940. That very year, many of the workers in the forming room reached retiring age, which facilitated the transition from hand-made to machine-made paper. Younger workers were selected and trained to tend the paper-making machine, and hand-moulding was gradually phased out. Nowadays all the paper needed by the Bank of Sweden is produced by machine.

Tumba's new sports ground was opened in 1941. This marked the fulfillment, after ten years' hard work, of a long-standing dream on the part of sporting youngsters at the papermill. Starting in the early 1930s, young chaps went out every spare evening to remove 7,000 cu. m. of clay, gravel and till to make way for the future sports ground. The football pitch was completed first, in the spring of 1936, and grass was sown on it that summer.

98

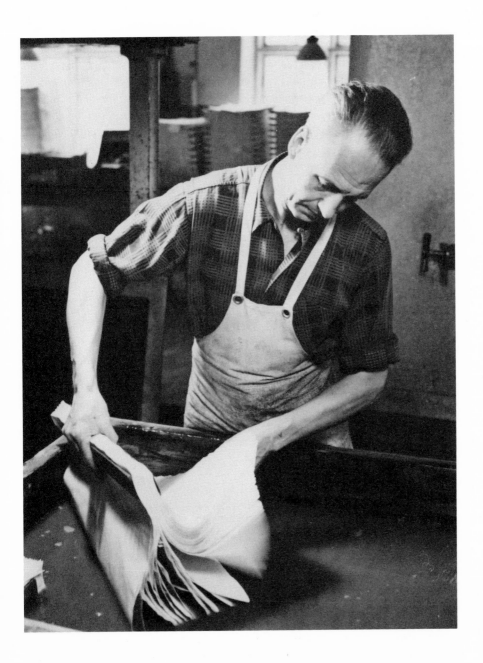

*Surface sizing by hand. Klippan Archives.* 99

But the sports ground also needed a running track and spaces for the throwing and jumping events, and I was in the team working on the latter. Work was soon able to begin, thanks to a grant from the Swedish football pools. As a result of a survey we ourselves had made of the material available, we believed that the members themselves could load and carry the infill for the tracks.

We also saved SEK 600 on infill by using the 200 cu. m. of spoil from our preparations for the football pitch. The many speakers at the opening ceremony paid tribute to the team spirit of the mill and the powers of initiative in which it resulted.

Production at Tumba Bruk continued without interruption during the war years, and I stayed in the manufacturing department, where I was made foreman in 1945. That year I went on a foreman's course lasting 200 hours, after which I was made a bank note inspector and shift foreman.

My job as shift foreman meant better earnings, and in 1945 Hulda and I bought a plot in Kyrkvärdsvägen where we started building a house in 1947. The house was finished in November 1948, and Hulda opened a food store on the ground floor. With me working shift, she had a hard time of it, but she kept the shop going until 1969.

In 1961 we performed the most remarkable assignment Tumba has ever had - the production of Wasa paper.

It was in March 1961 that the old warship Wasa was recovered from the waters in Stockholm where she had foundered in 1628. The excavation has revealed among other things the ship's anchor cable, which was made of hemp. The contract with the rope- maker, still extant, provided for four lengths of cable measuring between 110-120 fathoms each.

A qualitative examination showed that a great deal of the cable was in a very bad state. Some parts of it actually had to be "poured" ashore, while others were in a somewhat better state of preservation and could be lifted in pieces of about 2 metres, using metal pipes. Other parts again were so strong and well-preserved that they could be salvaged and conserved intact. The reason for these particular parts being in such a good state of preservation was thought to be the heavy concentration of tar and iron which they contained, which in turn was put down to the ship's own store of tar having flowed out over the cable.

In a normal excavation, the most heavily attacked pieces would not have been saved, but due to the special value attached to all finds on board the Wasa, the conservation committee decided that this material too was to be saved.

When the hawser was salvaged in July 1961, about 12 tons were brought up which were impossible to conserve. This contained various concentrations of impurities, e.g. water, tar, clay, rust and wood. The impurities amounted to some 75 per cent of the entire material.

*Tumba's first paper-making machine, an Escher Wyss. Installed and commissioned in the autumn of 1939. Tumba Bruk Archives.*

MAKING
PAPER

*Georg Anzelius and Harry Ericsson*
*checking the quality of the Wasa paper.*

The remainder comprised fibres of vegetable origin.

The fate of the unconservable cable was a problem until Lars Barkman, the chief conservator, suggested purifying the hemp and turning it into paper. And so Tumba was asked if it would be interested in taking part in a project of this kind.

The first we saw of the intended raw material was a sack of thick, muddy detritus. The contents were sent straight to the laboratory and rinsed in running water for several days. A vast quantity of coarse impurities, such as wood chips and tar, were removed in this way.

Once the laboratory staff had sifted out the fibrous material and put it through the beater, I produced a few trial sheets using a small mould in the laboratory sink. The dry sheets turned out a beautiful honey clolour with darker specks.

These trial sheets were approved by the Wasa Committee, and Tumba received an order for the paper. The format was put at more or less twice A4 and the colour was not to be paler than that of the specimen supplied. The paper was to be sized, so that it could be written on with ordinary Indian ink. The Committee also chose a watermark, consisting of a sheaf ("vase" in Swedish - the emblem of the Vasa dynasty), the letters "WASA" and the date "1628". The paper was to be made by hand, it was to be untrimmed and its grammage was to be about 90 g/m2.

The Wasa Committee's conservation department had the anchor cable roughly cleaned, toluol-extracted, fine-washed and fine- cleaned before it was sent to Tumba Bruk. At the mill itself, the following 16 operations had to be performed before the paper was ready for delivery:

1. Boiling.
2. Washing in water in the half-stuff beater.
3. Chlorine bleaching.
4. Alkali washing.
5. After-bleaching in sodium hypochlorite.
6. Washing in water.
7. Neutralisation of chlorine residues.
8. De-watering, removal from the beater.
9. Beating in the second-stuff engine and addition of rosin size, animal glue and alum.
10. Hand-forming of sheets.
11. Pressing of sheets in a hydraulic press.
12. Drying on a hot cylinder.
13. Finishing and sorting into whole and half sheets, tearing of half-sheets.
14. Pressing.
15. Folding of whole sheets.
16. Counting and packing, 100 sheets per packet.

Once stage 8 had been completed, it was time for the pulp to be put into the whole-stuff beater. To about 30 kg of Wasa cable were added 2 litres of Bevoid NX Special, 2 kg alum and 3 litres of animal glue. When the bleached cable had been charged, it was first beaten and mixed for 15 minutes, with the roll and bedplate just touching. The roll was then tightened 10 turns. The pulp was then beaten like this for 30 minutes, after which the roll was tightened one turn every 30 minutes during 4-5 hours' beating, finishing up with 15 minutes' whipping.

The first consignment after the test runs was delivered on 6th July 1962 and was a great success for the Wasa Committee. In spite of the high price, Americans almost fought over it. Some of them were prepared to pay up to 200 dollars a sheet. Altogether some 30,000 sheets were delivered and the proceeds to the Wasa Committee amounted to almost three million Swedish kronor. This was not bad going, considering that the anchor cable was past conservation. Instead it was converted into a highly exclusive and very beautiful paper which, moreover, had the great advantage of being saleable, which is quite out of the question where other museum exhibits are concerned.

It was about this time that the idea germinated of establishing a paper museum at the mill. True, there was already a small museum in the Bell House, but there was not room enough there to demonstrate the making of paper by hand. It was therefore decided to restore one of the oldest buildings at the mill, the former laundry, workers' quarters and paint shop, dating from 1763. This old building, with walls one-and-a-halfmetres thick, had stood up well to the ravages of time, and a vat room was fitted out on the ground floor, fully equipped with a hydraulic press and all the rest of it.

It was Harry Ericsson, a plant engineer at the mill who built up the museum and transferred to it all the old things from the Bell House and factory. On the top floor of the museum an instructional exhibition was arranged where visitors can trace the production of hand-made paper from rag cutting to finished sheet. A smaller room on the ground floor contains specimens of Tumba products, many different watermarks, a showcase with the famous hawser material from the Wasa, and a lot of other things besides. The museum was opened to the general public in 1968.

*Hand-forming paper at the Tumba Paper Museum. Left: The mould, with deckle attached, is dipped into the pulp. Right: Then it is raised horizontally.*

*Left: Shaking the pulp on the mould - the movement known as the vatman's stroke'. Right: The deckle is removed and the mould taken over by the couchers.*

In addition to activities at the museum, with demonstrations for school classes and other groups of interested visitors, we also take part every year in the Christmas market at Skansen, the open-air museum in Stockholm, together with an association called the Vat Team, which was founded in 1952 and consists of hand-made paper-makers from Tumba. The purpose of this organisation is to pass on the tradition of paper-making by hand to new generations.

The great interest aroused by hand-made paper is also reflected by our co-operation with the National College of Fine Arts, the National College of Art and Design, the Royal Institute of Technology and the Royal Swedish Academy of Fine Arts. One could not wish for more fascinated and interested pupils, and I am sure that the art of making paper by hand will be kept alive by artists, even when all paper-making by hand has been discontinued at the papermills.

*Left: The mould is placed at right angles to the wet felt. Right: The wet sheet is pressed onto the felt in a rolling movement.*

*Left: The mould is lifted off and the wet sheet stays put. Right: The deckle is put back over the mould and the procedure begins all over again.*

# PAPER PULP

**From raw fibre and rags to groundwood and chemistry**

The first paper-makers in China probably made their pulp from silk, both woven and unwoven. Later they came to use raw materials like hemp, mulberry bark, bamboo and various other fibres. The rags or bark, then, had to be beaten into a pulp which could then be mixed with water.

The original Chinese method was based on putting the raw material in a stone mortar, pouring water over it and pounding it until the fibres were separated. Pounding was done with stampers or mallets, a technique still in use in Nepal.

To this day, the Japanese pound their raw fibres from kozo, gampi and mitsumata by hand if a particularly long-fibred pulp is needed. But they use a wooden mallet, not a mortar.

The Chinese elaborated their art of paper-making and, by the 8th century, were selling quite considerable quantities to neighbouring countries further west, especially to the Arabs. The latter at this time were producing a great deal of papyrus, which was used in the capital of the Khalifate both under the Ummayads in Damascus (658-750) and under the Abbasids in Baghdad (from 750 onwards). The Arabs called the papyrus plant *bardî,* and one also comes across the name *fafîr* or *babîr,* which is an Arabisation of the Egypto-Greek papyros.

Parchment was used comparatively seldom, presumably for reasons of expense. It was mainly reserved for fine copies of the Koran.

At the beginning of the 8th century the Arabs conquered Transoxania and other eastern provinces, which brought them to the borders of China. During a conflict between two Turkish princes, one side formed an alliance with the Chinese, with the result that both came under Chinese rule. This prompted the victorious Arabs to intervene, and in 751 the Turkish princes were crushingly defeated near Samarkand.

The numerous captives included not only Turks but also Chinese, some of whom were paper-makers and revealed their art in exchange for their release.

Since the paper-mulberry tree, whose bark was used in China for making pulp, does not grow in these regions, a substitute had to be found. The fine fibres in linen rags then began to be used, and this is how the first linen rag paper came to be made.

The Arabs called this paper *kâghid,* a name which they borrowed from the Persian-speaking population in Samarkand, who in turn had got it from the Chinese *kog-dz',* "paper from the bark of the paper-mulberry tree".

*Pounding the raw material in a stone mortar.*

This paper was made from fibres of linen or hemp, though not directly but from rags, old cordage etc., in which the fibres had already undergone a certain amount of processing. The first thing to be done was to loosen all the knots and make the odds and ends of rope softer by combing them. The next step was to dissolve the fibrous material further by fermentation or by immersing it in lime solution.

For fermentation, the rags were cut up, put into piles, soaked in water and left to rot. The lime-water method involved putting the rags in a solution of lime, kneading them with the hands and then bleaching them in the sun. After this the rags were cut up into smaller pieces and rinsed in clean water. When all the lime had been rinsed away, the moist pulp was pounded in a mortar or ground between millstones until it was finally and evenly macerated.

The Arab method of pounding rags to pulp in mortars was, unquestionably, copied from the Chinese. During a later phase in the history of paper-making, both Chinese and Arabs began using a kind of treadle-operated hammer or stamper. This consisted of a long log with a heavy club fastened at one end. The actual tiltbar was balanced on the seesaw principle, and the paper-maker would depress the light end and then let the club end drop into a trough containing a

*Left: Paper-mulberry tree (Broussonetia papyrifera). The commonest bast raw material in Eastern Asia. Centre: Gampi (Wikstroemia canescens), which the Japanese consider the finest raw material for hand-made paper manufacturing. Right: Mitsumata (Edgeworthia papyrifera), a raw material for thin, beautiful sheets.*

mixture of raw material and water. This method was a great deal more efficient and did not demand so much labour as the earlier mortar-pounding method.

From China the art of paper-making spread eastward as well, across Korea to Japan. Unlike other paper-making countries, the Japanese have seldom used recycled materials such as rags. Their hand-made paper, *washi*, is made exclusively of fibres taken direct from bast. The three main types of raw material are called *kozo, gampi* and *mitsumata*.

The Japanese look on gampi as the king of paper-making materials, and gampi paper is considered unsurpassable in terms of both beauty and strength. The kozo fibre is somewhat coarser but immensely strong, while mitsumata gives a somewhat weaker paper. Japanese paper-makers refer to these three as a trinity, with gampi as the king, kozo the strong prince and mitsumata the beautiful but somewhat frailer princess.

Since the Japanese method of producing pulp is much the same today as it was a thousand years ago, the procedure is easily described. About 80 per cent of hand-made Japanese paper is based on kozo fibres, because kozo can be grown whereas gampi and mitsumata only occur in the wild state. The same pulp-making method, however, is applied to all three of these species.

The kozo is a perennial deciduous tree of the *Moracaeae* family and occurs in both the wild and cultivated states. The wild tree can grow up to six metres high, while the cultivated variety is hardly allowed to grow higher than between 1.5

111

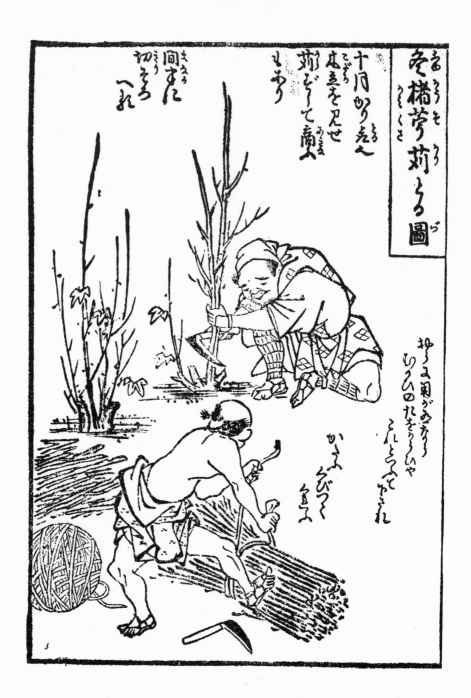

112     *Harvesting kozo. From Kamisuki Chohoki, 1798.*

 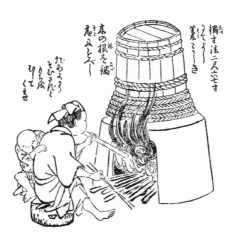

*Steaming the kozo sticks to ease off the bast. Left: Photograph 1985. Right: From Kamisuki Chohoki, 1798.*

and 1.8 metres, because the annual shoots are cut off every year at the root, which impedes growth. The kozo tree belongs to the same family as the mulberry, and these trees are so similar in appearance that, if one sees them growing in the same fields, it is hard to tell the difference.

Gampi (Wikstroemia canescens) is a wild shrub from which the Japanese have been making pulp since the 9th century. They were probably the first to discover its advantages.

Mitsumata (Edgeworthia papyrifera) as a paper fibre has a less well-known ancestry, but there is a document from 1597 in which a paper-making family requests permission to collect mitsumata bark. The plant belongs to the *Thymelaeceae* family.

Kozo has been cultivated in Japan for centuries. Seedlings are planted in spring on southward slopes. Roughly nine seedlings are planted every two square metres, regularly spaced so as to give them equal supplies of air and sunshine and make them easy to manure. The first year's growth is not thick enough for paper-making, but shoots from the second year onwards can be used. The shrub - it is in fact more like a small tree - puts out more than eight new shoots a year. After twenty years, however, the tree is too old and a new seedling has to be planted.

Kozo shoots are generally cut in autumn, after the trees have shed their leaves. The kozo bark cut during the autumn is called *akikawa,* "autumn bark", while that cut in February and March is called *harukawa,* "spring bark".

113

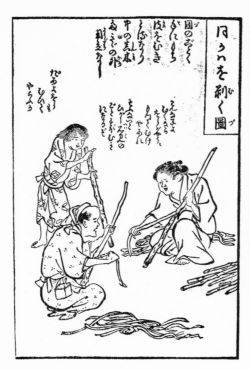

*Left: Stripping the bast from the kozo sticks. Kamisuki Chohoki, 1798. Right: A bundle of kozo bast, 1985.*

The autumn bark is considered to yield a better paper. The shoots are cut in lengths of about half a metre and tied into handy bundles. To steam off the bark from the shoots, the bundles, packed closely together, are placed on top of a large cauldron of water which is heated over a wood fire. The bundles are put on a grid so that the steam can rise freely. They are then covered with an inverted wooden cauldron.

Steaming takes about two hours, and after this the wooden cauldron is removed and the kozo bundles doused with water. After this "kozo sauna", the bark can be stripped from the sticks without difficulty. This is done while the shoots are still steaming hot, and little exertion is required. One simply twists the coarse end of the shoot, whereupon the bark cracks and can then be pulled off. It takes a lot of people - about 15 - to strip one batch of kozo before the next emerges from the "sauna". The bark is then put into bunches and hung up to dry. Drying usually takes three days, but in dry, windy weather, one day can be sufficient.

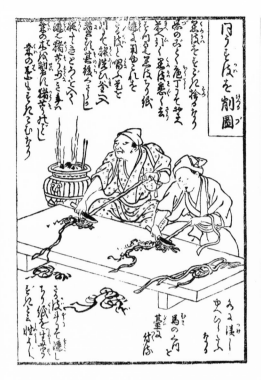

*Scraping the outer and inner bark apart. Left: 1798. Right: 1985.*

The dry bark is then stored as either black or white bark. If the pulp is to be made into writing paper, the black bark has to be turned into white first. Storage has left the bark dry and pretty hard, so first of all it has to be soaked in water so that the outer, black bark can be stripped from the white bark on the inside. This is done with a scraper knife, great care being taken to ensure that none of the black bark remains.

One ancient method was to put the bundles of bark on a flat stone in running water and trample them until the outer bark had been separated from the inner bark.

After the two layers of bark have been segregated, the white bark is carefully washed in running water to remove any specks that may still be adhering, and it is then put to dry in the sun for a few days.

When the time comes for making pulp, the bark is put to soak overnight, and the next day it is boiled in a large cauldron. When the bark has been on the boil for a while, lye of ash is stirred into it. The bark is then kept boiling, and stirred

115

*Beating pulp by hand with wooden beaters. From Kamisuki Chohoki.*

all the time, until it is so soft that it can be crushed between the fingers. The fire is then put out, a close-fitting lid is put on and everything is left to steam for several hours. By now the water has been stained dark brown, which shows that the chemical components of the bark, e.g. starch, fat and tanning agents, have been dissolved.

The cauldrons used for kozo boiling vary in size from about 100 to 200 litres. During the past 50 years, soda has begun to be used instead of ash, because its alkali residues are more easily removed after boiling. Soda is economical to use and, properly treated, the fibres are of great purity and the finished pulp drains more easily.

Caustic soda, which is a good deal stronger, can also be used, but this has the disadvantage of weakening the fibre. Strength, in other words, is sacrificed for purity.

Boiling leaves the bark a pale yellowish brown, and for absolutely white paper the bark has to be bleached. This, of course, can be done with chemicals, but to really bring out the natural beauty of hand-made paper, the Japanese feel that one has to use *kawazarashi,* the technique of bleaching bark in running water.

116

This is a primitive method but an ideal one, because it not only leaves the fibre intact but also brings out its strength and natural gloss. It is very much thanks to this method that certain hand-made Japanese papers have retained their beauty for more than a thousand years.

The bark is carefully rinsed after boiling to eliminate all impurities and alkali residues, after which it is deposited in a stream with flowing water and no large stones on the bed. A small dam is built round the selected spot, to divert any litter. A channel is then made to admit pure, clear water, and the water in this small pond is occasionally replaced, until the bark is bleached and washed absolutely white.

The importance of pure, clear water in Japanese paper-making cannot be overstated. Paper-makers prefer the ice-cold water - about 5°C - which flows in mountain streams during the months of winter. Most paper-makers, in fact, prefer working in winter, because the thickener, *neri,* used in the pulp goes furthest at low temperatures. The bacteria that can bear the effect of neri are happiest in warmer water, and so work in summertime is less remunerative.

Putting the finishing touches to the boiled and, perhaps, bleached bast is called *chiri-tori,* which means the removal of impurities. Even after boiling, bleaching and rinsing, some foreign particles may still remain, and if these are not removed there will be flaws in the finished sheet.

Cleaning is a laborious task, performed with the fibres floating about in icy water. One strand at a time is scrutinised before being passed for further progress in the paper-making process. All flecks of mould, patches of frost damage etc. are removed. If the impurities are too small to be picked out with the fingers, tweezers are used instead.

After all impurities have been removed, the bast, now snow white, is pressed into balls about the size of a melon.

Nowadays this classical method of turning bast into hand-made paper in keeping with Japanese tradition is only used for really fine, expensive paper. All the different operations have survived, but some of them are done by machinery. All fibres, for example, used to be crushed manually, with mallets. The pressed ball of fibre would be put on a hard slab of wood or a flat stone and the fibres crushed with a mallet or beater. This work was kept up until the fibres had been completely separated - a time-consuming task which, nowadays, is done mechanically, either in a stamping mill or in a beater of Japanese or European design.

Both stampers and hollanders will be described more closely further on in this book, but the Japanese have their own very special type of beater, called the *naginata.* This uses blades, like old-fashioned halberds, attached to a rotary shaft. Like the European hollander, it has a midfeather (middle partition), but

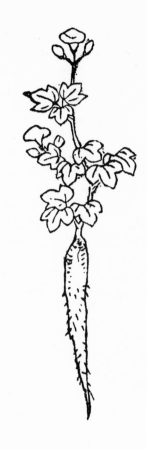

*Left: Japanese naginata beater. Drawing by Richard Flavin. From Barrett. Right: Tororo-aoi, the plant from which neri is made. From Kamisuki Chohoki.*

it has no backfall (return ramp), i.e. the ridge in the base of the hollander in front of the rotating knives, to keep the pulp from moving backwards.

Stamped bast pulp is run in the naginata together with water for one to twenty minutes, depending on the fibre quality required. Whatever his manufacturing method, the Japanese paper-maker takes care never to grind or beat the pulp longer than necessary. This is partly due to the traditional view that the precious fibre should not be subjected to more violence than is necessary, and then there is also a technical reason: the Japanese process requires very free, long fibres.

Another of the Japanese paper-maker's essentials is neri. This is a general term for a viscous agent added to the water in the vat to make it "thicker" and in this way to help in the forming of the sheet. Neri is related to the verb *nebaru*, meaning "to become sticky or viscous".

Generally speaking, the paper-maker will produce his own neri. It is extracted from a root called *tororo-aoi* in Japanese and *Abelmoschus manihot* Medikus or *Hibiscus manihot* in Latin.

The roots are dried in the shade and stored in a dry, airy place. The length of the root varies from 20 to 30 cm and it contains a great deal of transparent, sticky material. To extract this fluid from the root, the root has to be pounded with a mallet after it has been soaked in water for about 24 hours. After pounding the macerated roots are put in a cotton bag and pressed so that the tororo secretion is separated from the roots.

This transparent secretion is one of the absolute necessities in Japanese paper-making as it makes the water in the vat "thicker". This curious ability helps to slow down the escape of water from the su, thus giving the paper-maker more time to form the sheet as he wants. Without tororo in the vat, the long fibres would sink to the bottom of the vat and form lumps, and the water would leave the su so quickly that only a raw fibre mat would be left which could not possibly be formed into anything.

Tororo is a curious fluid in many ways. Even if it is often referred to as a mucilage, it has no glueing effects on the water. It does not feel sticky either. The best description of tororo - "more like thick water, clear and fluid and without any adhesive feeling" - comes in "Japanese Papermaking" by Timothy Barrett.

But tororo is not the only secret of Japanese paper-making. If a sheet is formed by the less vigorous western method of *tamezuki,* the result will be a thick, rather irregular sheet. The thin, extraordinarily uniform sheets can only be achieved by the somewhat more vigorous Japanese method of tossing the stock to and fro on the mould.

There is no dosage table for tororo. The paper-maker makes a rough appraisal of temperature, stock consistency, the intended thickness of the sheets and the state of the tororo, and he judges accordingly how much tororo he needs to pour into the vat. After a few sheets he knows whether drainage and dispersion are correct.

After this oriental excursion we can turn westward again. From Samarkand, the art of paper-making travelled further west, eventually reaching Spain and Italy by way of Egypt. This journey took about 400 years to complete.

The Arabs carried the art of paper-making further from China to the western world, but they did not take it any further in a technical sense. Their contribution, apart from the purely historical and civilising aspect, was that they began using rags as the raw material. They still retained the manufacturing methods which the Chinese had taught them; in other words, the rags were macerated in stone mortars or tiltbar stampers, as described earlier.

An Arab story, however, mentioned grindstones of some kind in Fez. These

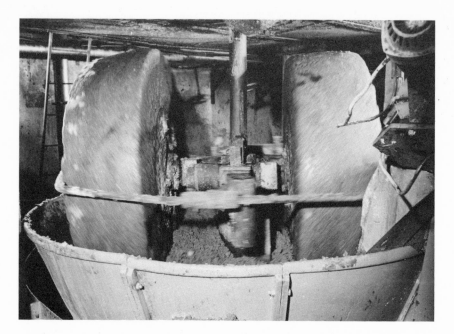

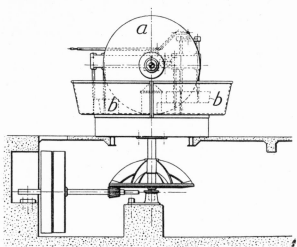

*The kollergang consists, basically, of two upright stones rotating in a trough filled with the material to be crushed. The stones rotate on a single shaft and act on a stone bed. The moistened raw material in the kollergang is crushed and torn apart between these grindstones in the bedstone. The kollergang was formerly used mainly for grinding pulped broke. In modern pulp manufacturing it is not used at all. Top: Kollergang in action at Frövifors Bruk. Photograph: Björn Nordien. Bottom: Structure of the kollergang. a. Stone, b. Scrapers. From Westergren.*

*Egyptian linen rag paper, 10th century A.D., with fabric traces and yarn threads. 2.5 x actual size. From Clemensson I.*

are supposed to have been used for making paper, and possibly the story refers to a kind of primitive kollergang or stone roll. This is made up of a circular stone trough in which a heavy stone wheel is made to rotate. The wheel crushes whatever is put in the bottom of the trough - rags, in this particular instance, but the same technique was used for grinding corn.

Dard Hunter found a similar mill in Korea which had been used for paper-making, and so one cannot discount the possibility of this method also having originated in the Orient.

Once the art of paper-making had arrived in Europe, it was not long before pulp-manufacturing methods were improved. A water-powered stamping mill was constructed in the Spanish city of Xàtiva in the mid-12th century. The main raw materials at that time were linen and cotton rags.

Well-worn rags of clothing and other fabric were wetted, pressed together into balls and left in piles. These were watered at regular intervals, to sustain the

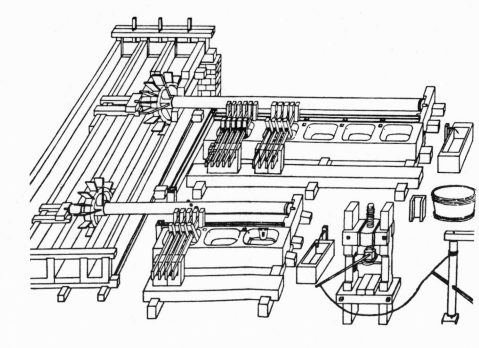

*Explanatory sketch of a water-powered stamping mill. The longest cam-shaft drives the stampers for five troughs, one shaft being fitted with five mallets. On the right can be seen a "Samson" and windlass.*

fermentation for the six or eight weeks during which the rags were left undisturbed. This was a very wasteful fermentation method, because at least a third of the material rotted and was unserviceable. The usable part, however, was easily processed in the stamping mill.

The stamping mills constructed in Spain consisted of a line of big wooden hammers or mallets attached to long timber beams. The opposite end of each beam was lifted by a cam, whereupon the heavy end, to which the mallet was fixed, dropped into a trough full of soaked rags. The trough could be carved out of stone or else part of an oak log. If it was made of wood, then usually it would be lined with iron or lead. These early European mills were water-powered, whereas their oriental counterparts were operated manually.

With the discovery of the rigid mould and deckle, the Europeans soon had a production sequence of their own for paper-making. An improved beating process, less prodigal of labour, was introduced and streams and rivers were made to power the innumerable papermills which now began to be built in the south of Europe.

*Papermill. From Theatrum Machinarum, 1662. The stamping mill in the foreground is water-powered.*

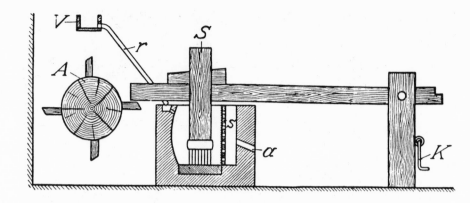

*Explanatory picture of a stamper. The cams on shaft A lift the iron-shod stamper, S, which is attached to a long beam. The hammer then drops into the trough, beating the mixture of rags and water in it. Water is supplied to the trough from the channel or leader V and the pipe r. It drains off through the screen s which prevents the pulp from escaping with the water through the outlet a. K is a hook to hold the hammer clear while the trough is being filled and emptied. From Westergren.*

After these basic improvements had been made to the process, though, it was some centuries before any further changes occurred. Well into the 15th century, the paper-making procedure was much the same as it had been in the 12th.

Larger stamping mills began to be built towards the 15th century. These were somewhat more advanced in terms of capacity and for the most part they worked by three stages. The first set of stampers had hammers shod with coarse iron teeth or spikes which tore the rags apart. This was done in a trough with a stream of fresh water washing the rags. The dirty water was led off through holes in the sides of the trough. Horsehair screens were fitted over the holes, to prevent the pulp from escaping.

This procedure replaced the more laborious method in use when paper-making first came to Europe.

Once the rags had been roughly cleaned and partly macerated, they were transferred to a new set of stampers, this time with somewhat finer hammers. Once again, though, a stream of fresh water was supplied to complete the washing process. This was either gravity-fed to the trough through wooden channels, or else it was pumped from the wheel driving the stampers.

For the most part the water was filtered in a series of reservoirs, with a fine-meshed iron grid at the outlet to trap any coarse particles.

The final treatment of the partly macerated rags, which were now comparable to a kind of half-stuff, took place in a third set of stampers, the hammers of which had no facing. This time there was no supply of running water, the reason being that by now the material was so macerated that a great deal of fibre would have escaped through the primitive strainers in the trough.

Ulman Stromer, first German paper-maker of note, who founded a papermill in Nuremberg in 1390, had a stamping mill in which 18 stampers were kept in motion by two cam-shafts. Stromer, who styled himself a merchant rather than a paper-maker, was quite a tyrannous employer. For example, his workers had to swear that they would not enter the service of any other paper-maker. Two Italians refused to take the oath, whereupon Stromer simply incarcerated them in his tower until they came to their senses.

In other parts of Europe too, there were stamping mills with many hammers per shaft. *Novo teatro di Machine et Edificii* by Vittorio Zonca (Padua 1607) contains the earliest known depiction of a water-powered stamping mill: an engraving of a mill with eight hammers operated by a single shaft.

In *Vollständige Mühlen Baukunst* by Leonhardt Christoph Sturm (Augsburg 1718) we read that it was quite common to have five stampers per trough and four troughs on each side of the shaft. In this way the shaft kept forty separate stampers in motion and the whole installation was powered by just one water-

*Dutch wind-powered papermill.*
*From la Lande.*

125

 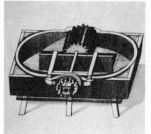 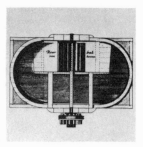

*Hollander with and without hood ("chapiter"), viewed from above. From von Natrus: Moolenboek, 1734.*

wheel. Stamping mills gradually developed from fairly rudimentary, clumsy machinery into fairly complicated devices. A large German papermill could have up to 25 troughs with four separate stampers in each. French papermills had three-five stampers per trough.

One of the distinguishing marks of paper made from stamping-mill pulp is its strength and durability. By studying old paper from this period one can tell, from the appearance of the fibres, the way in which they were reduced to pulp. The stampers did not cut or lacerate the fibres. Instead the raw material tended to be rubbed and frayed apart, and in this way it retained more of its original strength.

The Dutch, not having access to water power, used windmills but found it more and more difficult to compete with the Germans, who had plenty of water power at their disposal. Towards the end of the 17th century this prompted the Dutch to design a new machine for macerating rags. The new design did not need so much power as the stamping mill, added to which it was more productive.

This machine, which has been known ever since as the hollander, after its country of origin, is still being used today by hand-made paper-makers. The name of the inventor, on the other hand, is lost in the mists of time. Eventually the idea of a knife-armed roll to macerate the raw material put the stamping mill right out of business. Due to its rougher treatment of the raw material, the hollander was capable of using washed rags cut up into small pieces, without any previous fermentation. This made for a more economical process, and it was not long before the invention spread to other countries.

Dr Hans Bockwitz, in his *Zur Kulturgeschischte des Papiers* (1939), maintains that the first German hollander was in use in 1712. This was in Bavaria, and Bockwitz writes that, in five or six hours, it could macerate the same amount of raw material as a stamping mill in one day.

It was formerly believed that the hollander was invented some time at the beginning of the 18th century, but the discovery of a little book by Johann

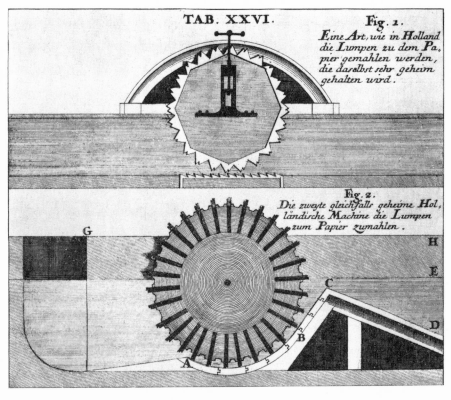

*Hollander. From Sturm: Vollständige Mühlen Baukunst, 1718.*

Joachim Becher published in Frankfurt 1682, made it necessary to revise this view. The book, entitled *Närrische Weissheit und Weise Narrheit: oder Ein Hundert so Politische als Physicalische, Mechanische und Merkantilische Concepten und Propositionen,* describes a new kind of papermill: "One does not know who conceived the art of paper-making, but it is a fine and wonderful invention. The commonest method is to fragment the raw material in a stamping mill, but in Holland I have seen a new kind of papermill which, instead of stampers, uses a roll and effortlessly grinds rags into pulp."

L.C. Sturm's *Vollständige Mühlen Baukunst,* already mentioned, contains an engraving of five hollanders powered by a single cog-wheel which also operates two millstones, so that both grain and rags were ground on the same premises.

Over the years the hollander has undergone several changes and has been specially adapted for a variety of purposes. One distinguishes between different kinds of hollander, depending on their uses. Basically they have their similarities, but they can be equipped for completely different functions.

127

When the hollander was in its infancy, the entire rag-beating procedure took place in a single operation, but as working methods grew more sophisticated, beating began to be divided into two procedures: half-stuff and whole-stuff beating, with bleaching in between.

The purpose of half-stuff beating is to turn the boiled and perhaps washed rags into a pulp of detached fibres which, at the next stage of production, will be easy for the bleaching fluid to get at and, in the final stage, will be a good input material for the end product.

Half-stuff beating is combined with washing of the fibre material when the half-stuff beater is fitted with a washing drum to strain off unwanted particles. After this beating and washing and the bleaching which follows it, the result is a semi-manufactured product called half-stuff.

Good half-stuff must be of perfectly even consistency - that is, it must not contain any unmacerated particles or fragments of fibre. A half-stuff beater is subjected to greater mechanical strain than the whole-stuff beater, due to the hardness of the unprocessed raw material, and so it is build more stoutly.

It is important for the hollander to have a good draught, i.e. for the pulp to circulate rapidly and steadily. Because the roll attracts the pulp in the hollander to it and below it, the level of the pulp declines in front of the roll. Since, on the other hand, it rises after the roll and backfall, the beater tub has to be of two different heights.

The first hollanders had a roll consisting of a turned block of oak or beech about 50 cm in diameter, fitted with the knives. As hollanders grew larger and larger, it gradually became difficult to find strong enough timber for the rolls, and so the knives eventually came to be attached to a cast-iron spider mounted on a shaft.

The roll also has to be adjustable to different heights above the bedplate, so that beating can be adapted to the gradual fragmentation of the pulp. At first the roll was raised by means of a setting screw on the roll shaft, but this way the roll was only raised on one side. Later another design was introduced which gave parallel elevation of the roll.

In order for a hollander to do its job properly, the bedplate has to be in first-rate condition. One of the difficulties in the earliest designs lay in achieving a good fit between roll and bedplate, and so the two were ground together by carefully lowering the roll against the bedplate while applying water and grinding sand.

As we have already seen, half-stuff used to be bleached in the half-stuff beater as well, but in the closing years of the 19th century this job began to be done in special bleachers instead. Bleaching in the half-stuff beater did not present much of a problem so long as the rags were fairly white and well-washed, but once the

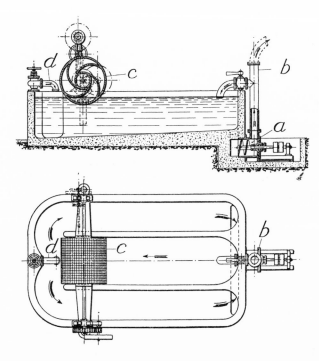

*Kron's bleacher.*
*a. Work-gear wheel,*
*b. Riser, c. Washing drum*
*d. Water supply*
*From Westergren*

*Paper pulp*

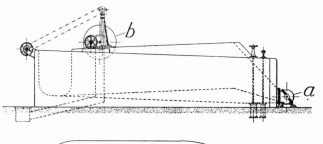

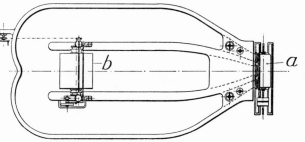

*Bellmer's bleacher.*
*a. Impeller housing,*
*b. Washing drum*
*From Westergren*

129

practice began also bleaching coarser and dyed rags, stronger bleaching liquid became necessary, and this caused rust damage to the knives. The rust worked loose from the knives and got into the pulp, causing ugly patches in the finished paper.

The bleacher was built on the same lines as the other hollanders, except that it was fitted with paddles instead of a roll. The backfall was made slightly lower and the bottom of the trough was rounded properly at the walls, to keep the pulp circulating as freely as possible.

There were also other bleacher models which differed from the ordinary type. In Kron's bleacher, for example, the pulp was impelled by a work-gear wheel in a metal cylinder outside the hollander trough.

The Bellmer bleacher was another version. This was a simple design in which the pulp was kept moving by a propeller, and in everyday speech it was known as a propeller hollander. The tank had one or two partitions, in the former case with a single-acting propeller (impeller) and in the latter case with a double-acting one to shift the pulp from the outer channels to the centre one.

*Two pictures from Dr Schäffer's book Versuche und Muster ohne alle Lumpen oder doch mit einem geringen Zusatze derselben Papier zu machen, Regensburg 1765-1771. In the picture on the right (potato experiments) his daughter painted a watercolour to show the durability of potato paper.*

Bleachers used chloride of lime, the active constituents of which were hypochlorites, partly decomposed by hydrolysis, because hypochlorous acid is extremely weak. The decomposed acid divides into more enduring compounds, namely hydrochloric acid and oxygen.

Hydrolytic equilibrium results from the hydrochloric acid being bound by the liberated alkali at the same time as the oxygen is consumed. Bleaching is due to the oxygen being released in these reactions and having a destructive effect on the compounds through which the fibre is coloured.

There were many factors involved in obtaining the best possible bleaching effect. The pulp had to be thoroughly washed. If the water in the bleacher was not clear, bleaching would take longer and more chloride of lime would be used, added to which the pulp would not turn out so white. A cloudy chloride of lime solution also meant poor bleaching and heavier consumption of chloride of lime.

Improved bleaching methods made it possible to use dyed and coarser rags, but the shortage of rags was a constant worry to paper-makers. A century and a half after the invention of the hollander, a solution had yet to be found which could improve the raw material situation. The ideas broached by the eminent

*Voith's grinder, at the Vienna Exhibition of 1873.*

physicist Réaumur of making pulp directly from wood, since the wasp was able to do so, had not led anywhere. Dr Schäffer's efforts to produce paper from all sorts of fibres were merely ridiculed and distrusted by paper-makers in the 1670s. Schäffer, however, was not out to find cheap, good substitutes for rags but rather to find substances which could be resorted to in a crisis. The worthy doctor, undismayed by the lack of vision shown by his contemporaries, persisted in his experiments and, indeed, succeeded in producing paper from poplar, pappus, wasps' nests, wood shavings, sawdust, beechwood and aspen wood, pine, osiers, straw, peat, asbestos and various other materials.

Dr Schäffer's experiments, then, almost coincided with the invention of the hollander, but more than 150 years were to pass before anybody solved the problem of finding a substitute for rags. The solution was designed by a master-weaver by the name of F.G. Keller in the German town of Kuhnheide, who had the idea of trying to defibre wood by grinding it instead of beating, which had been the method practised by Schäffer.

Eventually his experiments became so successful that he was able, in association with the paper-maker Heinrich Voelter in Heidenheim, to construct the first grinder. Keller's earlier experiments had been performed on a hand-powered grindstone, and indeed with very primitive equipment all round, but he realised that, if anything was to come of his idea, a machine would have to be constructed.

Beginning his experiments in about 1840, by 1844 he had reached the stage of producing about 100 kg groundwood pulp and getting it tested at a papermill. It was found that acceptable paper could be made from a mixture of 40 per cent rags and 60 per cent Keller's groundwood pulp.

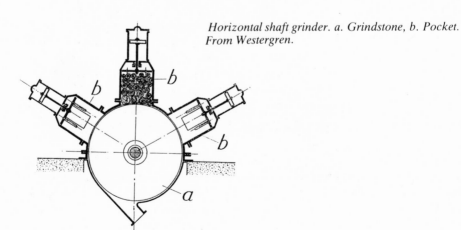

*Horizontal shaft grinder. a. Grindstone, b. Pocket. From Westergren.*

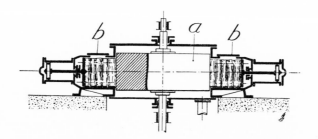

*Vertical shaft grinder. a. Grindstone, b. Pocket. From Westergren.*

Designing an efficient grinder took time, however. Voelter, Keller's partner, improved the grinder by several stages, but was still a long way short of a grinding machine which could be used in factory conditions. Voelter joined forces with the small engineering firm of J.M. Voith in Heidenheim to improve the grinding machine. After several years' experimentation, this workshop was able to improve the design of grinding machines sufficiently for them to be used commercially, and in time the company expanded into a worldwide supplier of grinding machinery.

J.M. Voith worked throughout the 1850s to elaborate the invention. In addition to developing the grinder, machinery had to be designed for divesting the groundwood pulp of splinters and coarse fragments, and also for thickening the stock.

Voith's grinding machines were a brilliant success at the 1867 World Exhibition in Paris. The method scored a complete break- through, and during the next five years Voith was commissioned to build 136 complete grinding plants in several countries. The invention of groundwood pulp could now be considered technically mature.

In his first experiments, Keller had used a vertical stone on a horizontal shaft, i.e. the same arrangement as an ordinary grindstone. Voith's first installations also had a vertical stone. These early grinders were called vertical grinders, as distinct from subsequent designs - horizontal grinders - in which the grindstone was positioned horizontally.

The principle of the grinding process was for the wood to be pressed against the grindstone by mechanical means. Counter-balanced levers provided the necessary compressive force. Before long it was realised that a great deal of compressive force was needed, and that the greater that force was, the better the output of groundwood would be. It was difficult, however, to evolve durable structures for these heavy pressures. Experiments were performed with screw presses, consisting of threaded spindles with shifter handles. As the wood was ground down, the presses had to be tightened by means of the handles.

# Pulpers for producing groundwood pulp.

### Fig. 1.
Horizontal pulper with 8 (4) weight and lever presses.

### Fig. 2.
Horizontal pulper with 8 (4) hydraulic presses and a pressure governor.

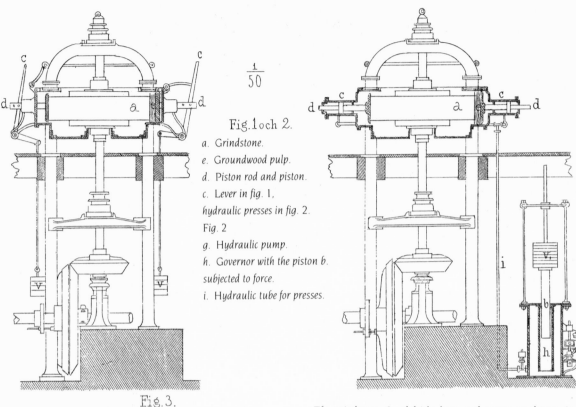

$$\frac{1}{50}$$

**Fig. 1 och 2.**

a. Grindstone.

e. Groundwood pulp.

d. Piston rod and piston.

c. Lever in fig. 1,

hydraulic presses in fig. 2.

Fig. 2

g. Hydraulic pump.

h. Governor with the piston b.

subjected to force.

i. Hydraulic tube for presses.

### Fig. 3.
Vertical pulper with 5 presses

The grindstone, 2, of fairly loose and coarse sandstone, is sharpened from time to time with a chisel or with a rotating, grooved steel roller held against it under pressure. The wood is ground under a stream of water and at an average pressure of 1 kg per 10 sq. cm. 300-400 litres of water per kg dry white pulp: nearly twice as much for pulp of boiled blocks.

A horizontal stone pulper, 1,300 mm in diameter and 360 mm high and with four press tables requires, at 190 r.p.m., 25 h.p. and produces about 260 kg spruce pulp in 24 hours. A pulper with a horizontal stone 1,700 mm in diameter and 530 mm high (the commonest size) and 8 press tables requires, at 150 r.p.m., about 90 h.p. and produces 1,000 kg grade pulp in 24 hours (as does a Wölters pulper with the corresponding power consumption).

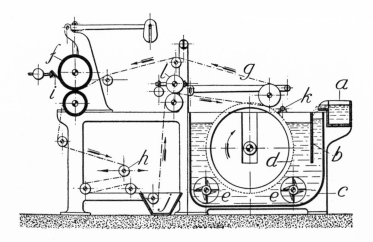

*Wet machine. a. Stock inlet, b. Froth board, c. Pulp chest, d. Cylinder, e. Agitator, f. Making roll, g. Felt, h. Felt roll, i. Doctor, j. Felt washer, k. Shower, l. Felt press. From Westergren.*

To guide the wood under compression against the stone, pockets were constructed in which the wood, cut to suitable lengths, was loaded. Pressure was then applied to this wood pile.

As grinding proceeded, abundant quantities of water were sprayed on to the grindstone, and this carried off the fibre produced by grinding. The concentration of fibre in the water was very low and this, coupled with none-too-efficient recovery machinery, caused a great deal of fibre to be wasted. Large quantities of pulp disappeared into the nearest watercourse, a form of pollution to which, at that time, nobody gave the slightest thought.

Voith's efforts to propagate his grinding technology all over the world resulted in Sweden, with its wealth of timber and water power, becoming one of the first countries to adopt the new method. The first grinding plant in Sweden was constructed at Trollhättan in 1857. This was the Öhnan groundwood pulp mill, which for ten years remained the only one of its kind in Sweden. It was not until 1866 that Klippan constructed a groundwood mill to meet its own demand for pulp.

Gradually more mills came to be built until, by the beginning of the 1870s, there were ten of them in Sweden. The rapid boom connected with the Franco-Prussian War of 1870-1871 inspired numerous manufacturers with confidence for the future, causing them to start up groundwood mills. This bonanza, however, soon gave way to a profound depression and, now that groundwood demand was satisfied, there followed a ten-year period of stagnation, without any new mills being built.

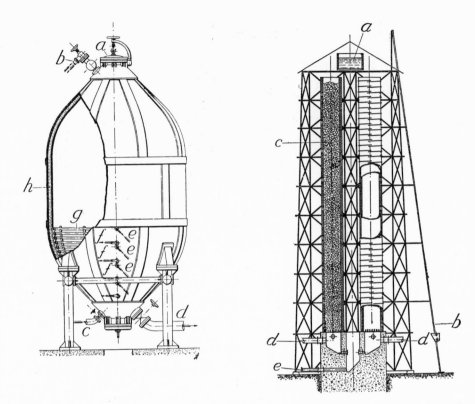

*Left: Mitscherlich's acid tower. a. Water tank, b. Limestone conveyor, c. Tower charged with limestone, d. Gas inlet, e. Pipe to acid tank. Right: Vertical sulphite digester. a. Manhole for charging chips, b. Piping for acid supply and de-gassing, c. Water supply, d. Discharge pipe, e. Steam supply, f. Steam outlet, g. Steam coils, h. Brick lining. From Westergren.*

Papermills all over the world began using groundwood pulp more and more as a result of the steep rise in newsprint production. The real upturn in Sweden came between 1885 and 1895. By the end of that period Sweden had no fewer than about 70 groundwood mills, some of them quite sizeable installations.

This new industry, which eventually was to be one of the foundation stones of Sweden's prosperity, was very much an export concern. Output at the end of the last century was about 120,000 tonnes per annum, of which 65 per cent was exported. This was a welcome source of revenue for a country as hard up for capital as Sweden.

Groundwood technology improved as time went on, and among other things the problem of fibre losses in the open system was overcome. The closed system evolved did not allow any water at all to escape into the surrounding environment.

136

*Centrifugal beaters are a different kind of maceration device from hollanders or kollergangs. These machines operate continuously, in that the pulp passes the knives or grinding discs once only, for cutting or shortening. The centrifugal beater, however, cannot replace the hollander entirely because a certain amount of greasiness is aimed for in the pulp. Consequently it is used mostly to supplement the hollander. B Drum with cutters. R Roll fitted with knives. A Pulp inlet. M, M, M Outlets. From Hofmann.*

From the mid-1890s onwards, however, there was a certain decline of interest in groundwood mills, the reason being that capital began to be invested in a completely new technology: the chemical sulphite method. *acid*

Alkali liquid had already been experimentally boiled together with straw, wood etc. at the end of the 18th century in an attempt to devise a material good enough for making pulp. When the *le Blanc soda method* became known at the beginning of the 19th century, there was a growth of interest in this kind of experimentation, but no method was ever devised for boiling wood with alkali liquor. The wood did not "dissolve" sufficiently, the reason being that, at atmospheric pressure, its boiling point was too low.

In the 1850s a Frenchman by the name of Mellier began boiling in an autoclave (under pressure). He boiled straw with 3 per cent soda lye in an enclosed vessel, obtaining a temperature of 150°. This released the substances holding together the cellulose fibres in the wood, and the residual product was a fibrous pulp which could be used for papermaking. Similar experiments were being conducted, at almost exactly the same time, by Watt in England and Burgess in America.

The world's first factory for manufacturing pulp by boiling wood was constructed in Pennsylvania, U.S.A., in 1860. The pulp produced was alkaline cellulose, a brown pulp in which sodium carbonate was added to the stock.

Alkali cellulose was the earliest of the wood-fibre pulps to be produced by boiling, but in time it was superseded by sulphate cellulose. This again was brown in colour and came to provide the raw material for all brown paper, e.g.

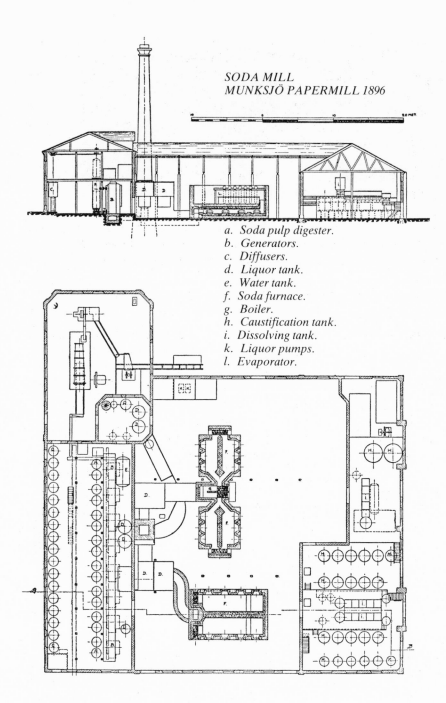

SODA MILL
MUNKSJÖ PAPERMILL 1896

a. Soda pulp digester.
b. Generators.
c. Diffusers.
d. Liquor tank.
e. Water tank.
f. Soda furnace.
g. Boiler.
h. Caustification tank.
i. Dissolving tank.
k. Liquor pumps.
l. Evaporator.

138    *Munksjö Soda Mill in the 1890s. From Molae Chartariae Suecanae.*

wrapping paper. Sulphate cellulose, or sulphate pulp as it is equally often termed, is produced by boiling spruce or pinewood with alkali liquor under pressure. The cooking liquor consists of sodium hydrate and sodium sulphide, and it is obtained recovering alkalite from the black liquor which remains after cooking. This is done by evaporating the black liquor to dryness and burning the dry substance. The inevitable loss of alkali is made up for by adding sodium sulphate (glauber salts) to the dry substance before it is burned, whereupon the sodium sulphate is reduced to sodium sulphide.

Five years after the first alkali cellulose plant had been built, another pulp mill was built outside Philadelphia. The chemist B.C. Tilghman experimented for some years at this mill and in 1867 took out an American patent for a method of "treating vegetable substances ... for the manufacture of ... paper pulp". Tilghman originally proposed digesting the wood in a lead-lined, cylindrical, rotating vessel, but progress was delayed by the immense mechanical difficulties involved.

Probably unaware of the Tilghman patents, the Swedish chemist C.D. Ekman started the first sulphate mill at Bergvik, Sweden, in 1873, using magnesium bisulphite solution as the cooking agent. He did his utmost to keep the method a secret, and it was not until 1881 that he took out a British patent for it. Ekman's mill was equipped with small rotating digesters heated indirectly by steam.

At about the same time as Ekman, the German chemist A. Mitscherlich was working on the same problems, and in 1878 he took out a German patent for his procedure. His method involved using a horizontal, stationary and cylindrical digester lined with brick and indirectly heated. The wood was cooked at low temperature and pressure, which made for quite a slow process.

There ensued a battle royal over the credit for inventing the sulphite method. In the end, Ekman was awarded priority for the invention, and he was indeed the first person to make practical use of the method.

New developments came thick and fast round about the 1870s. The Belgian E. Solvay, for example, perfected his ammonium soda method and began competing with the old soda method of le Blanc. This soon caused the world market price of soda to plummet, which in turn led to a cooling off of interest in Tilghman's sulphite cellulose method, soda being the heaviest item of expenditure where soda pulp was concerned.

Instead it was hoped that soda pulp would be the thing of the future. The Swedes took a close interest in all industrial facilities opened on the Continent, especially those using raw materials abundantly available in Sweden. Timber was a raw material of this kind admirably suited to "sunrise industries" in Sweden, and the first soda pulp plant was opened at Delary, Småland, in 1871. This pioneering venture was organised by the energetic Count S. Lewenhaupt, who founded another mill at Värmbol, Södermanland, soon afterwards.

The technical know-how used by the Swedish pioneering enterprises originally came from Britain, and before long it became clear that there was room for improvement. The founders of the soda pulp industry had to put in a great deal of hard work before even the main outlines of the technical methods were established. Eventually these labours resulted in a Swedish system of soda pulp production, and from very modest beginnings there gradually emerged a big industry - the sulphate industry - which achieved a position of international leadership. The engineers who created that industry learned the first lessons of their trade at the pioneering facilities in Sweden.

The small mill at Delary produced only two tons of pulp a day with its single digester. After a fire in 1875 the mill was rebuilt and three digesters installed. This raised capacity to five tons daily and, eventually, to twelve.

One of the difficulties in production was the recovery of alkalite. By a coincidence, a manufacturer called G.A. Engelbrektson got to know an engineer, Alvar Müntzing, who at that time was working at the Inedal Sugar-Beet Mill. Engelbrektson described his problems and Müntzing suggested a trial of the diffusion method used in the raw sugar industry.

Under Müntzing's direction, experiments were conducted at Engelbrektson's soda mill, washing soda pulp in a battery made up of paraffin drums joined together with piping. Müntzing's idea soon proved correct, because the pulp was washed without the liquor being significantly diluted. The mill was immediately altered in keeping with Müntzing's proposals and so the diffusion process became part of the soda pulp and succeeding sulphate pulp industry. This Swedish achievement is rightly regarded as a milestone in the history of the pulp industry, and Müntzing later changed horses, devoting himself exclusively to the chemical pulp industry. The diffusion process was soon introduced at all soda mills in Sweden, and the biggest difficulty connected with alkali recovery had now been overcome.

Müntzing also tried boiling wood in a mixture of sodium lye and sodium sulphide, a solution already suggested by J.L. Jullion in 1855, but his experiments did not have any lasting practical results. Instead it was a German, C.F. Dahl, who was to pioneer the sulphate digestion method. This method had the advantage of greatly increasing the chemical pulp yield, at the same time as the pulp was both stronger and of better quality. Sodium sulphide, moreover, was cheaper than soda. The soda pulp mills quickly changed to the new method, which dispelled the difficulties of pulp yield and brought down operating costs. The soda pulp method was now entirely superseded by the sulphate pulp method.

Müntzing continued his experiments with sulphate digestion at the Munksjö

*Apparatus from Ekman's laboratory at Northfleet. National Museum of Technology, Stockholm.*

sulphate pulp mill, and in 1885 he devised a new digestion method which was to make all the difference to the future of the industry. With this new method he produced "kraft pulp" for making a very strong paper, known as kraft paper. This was a brown wrapping paper which eventually became a very important export item.

C.D. Ekman's sulphide method made it possible to produce a white pulp without using rags. Ekman carried out his experiments in 1872 at Bergvik, which at that time belonged to a British company. By using an acid magnesium sulphide solution, he achieved good results by cooking the wood under pressure. Bergvik at that time was a small mill producing brown fibre pulp; the wood was cooked in water and then fragmented. Ekman tried to bleach this fibre with sulphite and this is what led him to experiment with sulphite digestion.

On receiving a specimen of the Ekman's pulp, the British company immediately decided to commence experiments in a small mill. That small mill, at Bergvik, was the first sulphite mill in the world and it began operating in 1874.

Ekman's "rival", Mitscherlich, used calcium bisulphite as a cooking liquid from the very beginning, realising as he did the economic necessity of using the cheapest possible base, i.e. lime. He started his experiments in small 1/2 $m^3$

141

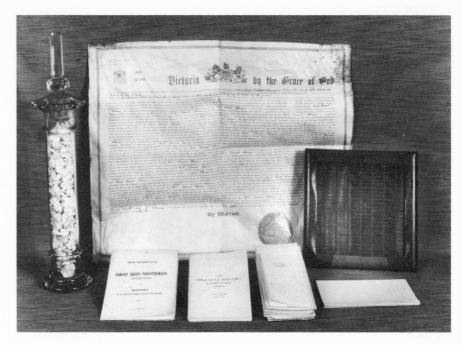

*Ekman's British patent; writings and paper specimens; paper mould for Ekman's private letter paper; sulphite pulp specimen from the 1897 Stockholm Exhibition. National Museum of Technology, Stockholm.*

digesters, later rising to 9 m³. He had far greater difficulty, however, in evolving a technique which could be upscaled into a factory. For one thing, gypsum formation in the digesters was a serious obstacle. Eventually Mitscherlich transferred his patents for the sulphite process to the German firm of Gebr. Vogel, which in 1880 built the first real sulphide mill, equipped with digesters of about 80 m³ capacity. A number of other Swedes besides Ekman were experimenting with sulphide digestion at the beginning of the 1870s. At Mölndal, near Göteborg, an industrialist by the name of D.O. Francke was running a papermill where, at the end of the 1870s, sulphite pulp was being produced on a technical scale. The cooking liquid used was calcium bisulphite, and this was presumably an international first. The mill did not really prosper, however, and so Francke and his engineer, C.W. Flodqvist, went to Germany and studied Mitscherlich's methods as upscaled by Gebr. Vogel. The experience thus gained was fed into a new mill, opened in 1882.

Sweden's first really big sulphite mill was built by V. Folin in Säffle, in 1883. Here, for the first time, the digesters were enclosed in brick walls. In his Swedish patent, Folin describes the brick-lining process by saying that the digester shell

*Billerud sulphite pulp mill in the 1890s. From Molae Chartariae Svecanae 1923.*

is coated with boiled coal tar or asphalt to which a thin sheet of lead is attached. A layer of refractory brick is then laid against the lead sheet.

The earlier technique was to line the insides of the digesters with lead, but this method proved to be untenable in the long run. Alternate heating and cooling caused the lead sheet to "creep" and buckle, and so the digesters had to be overhauled after every charge. Most mills later adopted brick-lined digesters instead.

Sulphite and sulphate pulp now conquered the world, and rag pulp came to be reserved for archive, bank note and document paper, which had to be extra durable.

Eventually rag pulp became almost impossible for the papermills to use. The costs of cleaning and sorting the rags and the increasing element of artificial fibres led the mills specialising in document paper production to substitute linters, a pure cotton product made from the short fibres surrounding the cotton seed. Most of the long fibres collected are used in the textile industry, e.g. to make ginned cotton. The shorter fibres are used for pulp and chemical products.

Linters used to be a waste product. The fibres were removed as far as possible from the valuable seeds used in the production of oil and feed substances. But today, cotton linters are important raw material for fine papermills. Simon Barcham Green, who ran Britain's last commercial hand-made papermill, closed

143

down in 1988, writes in the journal Fine Print that an average delivery of linters to a pulp mill contains 73 per cent cellulose, 11 per cent seed hulls, 8 per cent water, 2 per cent fat and wax, 2 per cent ash, 2 per cent protein, 1 per cent pectins and 1 per cent sand. After chemical purification and cleaning, a pulp can be produced consisting of 99 per cent cellulose, of which 99.7 per cent is alpha cellulose and a mere 0.3 per cent the less desirable beta and gamma cellulose. Cotton fibre is not very strong and is preferably used together with other fibres, unless a particularly soft paper is required, e.g. for etchings.

*Cotton linters,* then, are practically the only material available for professional production of hand-made fine paper. Cotton is the purest form of cellulose and its fibre the principal raw material for hand-made paper production in the west. Other fibres, however, can also be used, even though they are practically unobtainable in economically viable quantities.

*Flax* is a bast fibre, just like kozo. The useful material, contained by the inner bark, equals only about 5% of the total mass of the plant. To get at the bast, the flax has to be pretreated by retting, i.e. bundles of flax are put in ditches where biological degradation softens the bark, making the bast underneath relatively easy to get at. Further chemical processing removes lignin and other non-cellulose material. The traditional method of bleaching linen was to soak it in running water and dry it in the sun, not unlike the methods used in the Orient. Paper made from pulp consisting entirely or partly of raw flax fibres has a hard, beautiful surface which makes it ideal for calligraphy and book wrappers.

*Manilla hemp* also known as *abaca,* is the strongest of the fibres used in western hand-made paper production and serves mainly as a reinforcement for cotton linters. Manilla hemp is derived from the bast fibres of various banana species.

*Hemp* and *jute* are other fibres which can be used for papermaking. Hemp was one of the first fibres to be used for paper in the Orient. Jute resembles it but is slightly coarser. These two fibres are very rarely used in western papermaking nowadays. The growing of hemp is prohibited in Sweden and most other western countries, due to its involvement in the illicit drug trade. Raw manilla hemp, hemp and jute contain quite considerable amounts of lignin which have to be removed by means of strong chemical agents. If the same methods were to be applied to cotton fibres, they would not be worth much afterwards, but these fibres are so strong that, even if they are damaged by rigorous chemical treatment, they still emerge stronger than cotton fibres.

*Rags* have been dealt with on so many previous occasions in this book that no further review is needed of the procedure for manufacturing rag pulp. It is perhaps worth pointing out, however, that a distinction has to be made between rag pulp and pulp made from raw fibres. The distinction tends to be obscured in the English-speaking world by linters frequently being sold as "100 per cent rag", whereas in actual fact the raw materials consist of spinning mill waste which never came anywhere near a loom. The word "rags", thus used, has come to denote cotton in any guise whatsoever.

As the worthy Dr Schäffer already pointed out at the end of the 18th century, paper can be made from most vegetable fibres. *Nettles, potatoes* and *iris leaves* are excellent raw materials for papermaking on a small scale, e.g. for art use.

*Waste paper* is an important input material for paper mills in their production of cardboard, millboard and other coarser kinds of paper which do not have to be particularly durable or light in colour. Waste paper, however, is of little interest to hand-made papermills. It consists mainly of newspaper which is made from wood pulp, is sensitive to light and embrittles rapidly, as most readers will already have noticed if they have ever left a newspaper in the sunshine for a few hours. It goes without saying that hand-made paper, with durability and lightness of colour as its principal raison d'être, cannot be based on recovery pulp. One exception to this rule is Tumba, which manufactures a pulp made from pulped-down bank notes, containing up to 90 per cent rag.

# THE PAPER MOULD

### The paper-makers principal tool

The hand-mould is the paper-maker's principal tool and the embodiment of 2,000 years of development in the art of papermaking. Other aids may vary from one age or country to another, but the mould is the one thing that no paper-maker can manage without. This is where the fibres of the pulp are matted together, the water drained off and the paper sheet fashioned.

No wonder, then, that generation upon generation of paper-makers have struggled to adapt the mould to their own requirements - requirements which have varied according to the material available and the production methods used. Due to its very nature and the treatment it is subjected to, the life of a paper-mould is quite short. The mould in daily use at a hand-made papermill will last little more than one year, and if sparingly used it will last for up to ten years. For this reason, old moulds are extremely rare and the oldest extant in Europe only date back to the end of the 18th century. We can only guess at the appearance of earlier Oriental and Occidental moulds. Our only possible means of investigating this topic is by studying old paper and the primitive working methods still being used, for example, in Nepal.

The first paper is supposed to have seen the light of day in southern China, where bark fabric, an unwoven material, was produced previously. The technique of making this material then spread eastwards to Indonesia and the countries of the Pacific and westwards to Africa. We may also assume that the floating mould was developed for making a product which resembled the bark fabric and not primarily for the production of writing materials. That theory is supported by the fact of the same raw material being used, e.g. the inner bark of the mulberry tree *(Broussonetia papyrifera)*, and by the large format of the paper.

The floating mould is still used in southern China, Thailand, Burma, Nepal and Bhutan, although in most cases it is more a tourist attraction than a matter of commercial papermaking. This mould consists of a rectangular wooden frame with a fabric underneath consisting of China grass or widely spaced cotton. The mould is floated in a pond or small stream, and the stock is then poured on to it and smoothed out manually. This kind of mould requires no deckle and no ribs for the fabric, as found in subsequent types of mould. On the other hand, one can sometimes see handles at the sides, for lifting the mould and putting it to dry in the sun or close to a fire. The floating mould, in other words, can only be used for making one sheet at a time. Unlike subsequent types of mould, the wet sheet

149

*Papermaking in Nepal. Soaking bast. All the pictures on these three pages were taken by Anna-Grethe Rischel.*

*Beating with a wooden mallet.*

*Mixing with water.*

*Pouring the stock on to the floating mould.*

*Levelling the stock in the mould.*

*Lifting the mould for the surplus water to run off.*

151

*The mould and newly formed
sheet are put to dry.*

cannot be couched on a felt, nor, as in Japan, can wet sheets be piled on top of each other. The wet sheet is far too delicate to be removed from the floating mould before it has dried, and so production is immensely time-consuming and requires a large number of moulds. One advantage of the method, though, is that it requires no vat and the paper-maker can use a minimum quantity of pulp.

In his *Ancient Paper of Nepal,* Jesper Trier describes the procedure for papermaking with floating moulds in eastern and central Nepal, at an altitude of between 1,800 and 3,500 metres along the caravan routes to Tibet. Most paper-makers have papermaking as their principal occupation and live for 3-10 months in the forest, while their relatives and families tend the farms and fields lower down. A temporary hut is built near a stream to accommodate the papermaking equipment as well as the paper-maker and his family. The production method is the simplest imaginable and the most primitive in existence, but it is pretty efficient. The sequence is as follows:

1. *Harvesting* bast. This is usually done by women and children.
2. *Soaking* the bast and removing black specks.
3. *Cooking* in ash lye (often with sap from Rhododendron arboreum leaves added). About 20 litres are required for every 2.5 kg bast.
4. *Beating* with a wooden mallet. This takes about 1.5-3 hours.
5. *Removal* of lumps from the bast.
6. *Dilution* with water (about 1/40, depending on weight) using an instrument rather like a bamboo butter churn.
7. *Pouring* the liquid stock into moulds consisting of a plain wooden frame with a coarse fabric fastened underneath it. The sheets measure about 40-50 cm x 60-65 cm. Each hut owns 20 or more moulds and produces about 200-500 sheets per day.
8. *Drying* round a big fire, turning the moulds through 180°. So much wood is used up that the paper-makers have to change locations every 3-4 months.

9. *Folding* the sheets. Bundles of 200 sheets are put together in a bigger bundle. The raw material consists entirely of Thymelaeceae fibres of different qualities. The commonest quality has a grammage of 15-20 $g/m^2$ and 2.5 kg bast yields about 250 sheets. The paper turns out brown or reddish brown in colour and it is thin, frequently containing both holes and pieces of bast. The best quality is less brown and has a grammage of 30-35 $g/m^2$. About 5 kg bast will make roughly 250 sheets. With Indian ink one can write on both sides of a sheet.

The floating mould may be admirably suited to primitive papermaking, but it is out of the question for meeting big demand for raw materials, such as occurred, for example, during the Han period (206 B.C. - 220 A.D.) in central China. Presumably the Chinese first encountered primitive papermaking in the neighbouring countries to the south, after which they improved the methods as demand increased. The technique of the mould was refined and for raw material the Chinese added hemp fibre to the traditional bark fibre. Thus originated the various oriental types of paper-mould which are still being used in the Far East.

A more poetical historiography tells us that the Chinese eunuch Ts'ai Lun hit on the idea of turning the macerated fibres of old clothes and suchlike into a material for writing on. The story goes that, contemplating a tub filled with old rags that had been beaten in a mortar, Ts'ai Lun observed loose fibres floating on the surface of the water. The problem now was to extract them from the tub and to make them felt together and then dry into a sheet. Nobody knows what the mould he invented looked like, but presumably it was of the floating variety, which meant that the stock was poured over it; but it may also have been of the kind which is dipped into a vat of macerated stock and then lifted up to dry.

One can only guess at the process whereby the floating mould developed into the flexible version still being used in the Far East. Presumably the process took several hundred years to evolve. The flexible mould consists of a wooden frame reinforced with ribs and either a loose or a fixed deckle. The ribs are often triangular, with the apex at the top, flush with the frame. Stretched across the frame is a screen, supported by the ribs, which replaces the coarse fabric of the floating mould.

The screen in this type of "laid" mould is made of parallel strips of bamboo or grass, close-packed and held together by silken threads or horsehair. Compared with a woven fabric, the advantage of a laid fabric is that the water will drain off more easily through the oblong slits than through the square holes of the weave. It is unknown from what craft the paper-makers of the Han period derived their moulds, but it is worth remembering that the Japanese employed the same technique in the straw mat used as a placemat. The finds made by Sir Aurel Stein and Sven Hedin in East Turkestan and Loulan also show that the paper they

153

found had been made on a laid mould with strips of bamboo. In his *Papermaking*, the eminent paper expert Dard Hunter takes the view that the first Chinese papers were made on a woven mould but that there are no finds to support this theory. No third century paper bearing the impression of a woven mould has been found, but Hunter still believes, on the strength of his own experiments, that the woven mould must have preceded the laid version. With this mould the stock was simply scooped up and poured out over the silk cloth, whereupon the mould was left to dry. Possibly this type of mould was only used for a short period, but long enough for the person who had invented the method to see that a very useful material could be made this way - a material excellently suited to calligraphy, an art which had previously been practised on silk and pieces of wood.

Although we know very little indeed about the earliest Chinese methods - the earliest descriptions date only from the 1590s and 1630s *(Pen t'sao kang mu, and T'ien kung k'ai wu)* - the Turkestan finds show that high-class paper was being produced in China very early on. Papers from the 3rd century B.C. to the 7th century A.D. are generally made of thoroughly macerated pulp, and are thin and translucent. Most of them are also sized and are easy to write on with a steel nib and ink.

Most paper from the early centuries of the Christian era is white, but from the 5th century onwards it requires a greyish shade. From the beginning of the 8th century onwards, quality rapidly deteriorates and paper becomes limp and thick and does not retain the ink at all well. The paper examined from the 10th century is inferior in quality to papers from the 6th and 7th. Many of the papers found in East Turkestan have been dyed after forming and drying. Yellow and brown shades are commonest, but blue also occurs. It is uncertain, however, whether the decline in the papermaker's art reflected by the finds in East Turkestan was really true of papermaking in China. A lot of the paper found in East Turkestan can very well have come from other parts of Asia, e.g. Tibet and neighbouring regions. At all events, the inferior quality displayed by certain paper finds can hardly have had anything to do with the technology of the mould.

The big change occurred when some genius discovered a method of detaching the wet sheet from the mould cloth without the mould having to dry first. The problem is that the newly formed sheet is very delicate and can hardly be touched without tearing. To make it come away from the mould covering, the latter has to be made of a material which is both soft and firm. The unknown inventor constructed a kind of matting consisting of thin, rounded bamboo strips laid side by side and sewn together with silk, linen, camel-hair, yak-hair or horsehair thread.

This kind of mould is called "laid" with reference to its structure of "laid" bamboo strips. Old European moulds too were made the same way, but using brass or copper strips.

This bamboo mat was placed on top of a frame with ribs. No deckle was used. Instead the entire implement was dipped into a vat full of stock. It was then lifted up again with a thin layer of stock covering the detachable bamboo mat. A thicker bamboo rod was sewn parallel to the more slender ones on two sides of the mould and the paper-maker held two separate rods against the other two sides; in this way the stock was prevented from flowing over the sides. When sufficient water had drained away from the wet sheet, the paper-maker couched the sheet on to an underlay. The word *couch* comes from the French *coucher*, to lay down; old French *colcher, culcher* from *colchier;* from the Latin *collocare*, to place. To do this he lifted the matting and the wet sheet clear of the wooden frame and placed it, with the sheet downwards, on a board. After this he lifted the matting clear with a "rolling" movement, leaving the wet sheet on the board.

One after another the sheets were then couched on top of each other. This was a tremendous advance compared with previous methods, whereby it was only possible to produce one sheet per mould and the sheet then had to dry before the procedure could be repeated. With the couching method, one paper-maker using one mould could do the work of a whole team.

Complicated as it may seem, this type of mould was the best and simplest method available, and right down to the present day, it is the foundation of the oriental art of papermaking. Even the most modern paper machine is based on the same principle.

An intermediate or transitional form between floating and laid mould is described by the paper expert E.G. Loeber in his *Paper mould and mouldmaker*. This mould consists of a wooden frame with triangular ribs and a mat made of grass sewn together with horse-hair, and two deckle sticks to keep the mat pressed against the frame when it is dipped into the stock.

Before dipping the mould, the paper-maker stirs the stock with a stick. He then waits a while for the coarser fibres and any lumps to sink to the bottom, but while the finer fibres are still floating on the surface he grasps the mould at both ends and secures the mat by clamping it with the two deckle sticks. After carefully stirring with the lower part of the mould, which is held vertically, he dips the mould into the stock and holds it horizontally beneath the surface until a sufficient amount of pulp has floated in and covered it. The mould is then left floating on the surface until the newly-formed sheet has drained. After this the paper-maker pushes the mould under the surface to dislodge the sheet from the mat; this simplifies the ensuing couching operation. Now the paper-maker lifts the mould and sheet out of the vat for a second dip, which adds another layer of pulp to the first. By shaking the mould carefully, he causes the two layers to mat together. The mould is then carefully lifted out of the vat in the horizontal position and propped up against a support so that the water will run off.

The grass mat and its wet sheet are then lifted off the frame with a rolling movement and deposited on a board, once again with the same rolling movement, until the entire sheet has been laid out. The mat is then carefully lifted clear, starting from one side.

The entire procedure of dipping and couching takes about four minutes and is repeated until a post of about 140-170 sheets has been completed. The whole bundle is then pressed and dried. The maximum daily output is something like 250 sheets. Although mould designs and methods of containing the wet stock within the bounds of the mat vary somewhat, the mat or screen and the entire working procedure are virtually identical.

It is impossible to say exactly when the Chinese began using laid bamboo moulds, but it must have been several centuries before the art of papermaking arrived in Europe. As far as we know, no watermark of any kind has been found in paper from the time of the first vegetable mat moulds. Perhaps early paper-makers never thought of this way of marking their paper, or else the bamboo strips were too inflexible for such an idea to be put into practice.

Quite certainly, however, the original type of laid mould invented by the Chinese served as a model for the "chain-lined" moulds which appeared subsequently, and nowhere in the world have more than small modifications been made to the basic design.

From China the art of papermaking spread to Korea. Korean hand-made paper has always had a very distinctive appearance, due to the moulds on which it was formed. Just like the Chinese laid mould, the Korean version consists of four separate parts: the frame, the laid-cover (i.e. the screen or mat) and two deckle sticks. Although bamboo is the material most commonly used for making Korean moulds, sometimes another species of tall grass has been used instead. In India too, grass has long been used for laid mould covers.

All hand-made Korean paper presents the same distinctive characteristics. The impressions of the "laid" bamboo strips run in the narrow direction between the sides of the mould, while the "chain lines" of the threads holding the strips together run the length of the mould. A Korean paper-maker also uses a frame above the vat, in which the mould is hung from a string attached to one end. The paper-maker forms the sheets by dipping the mould, sideways on first, holding it by one end. He then distributes the stock by shaking and swinging the mould. The laid cover is secured between two wooden frames, and when forming is complete the two halves of the frame are detached and the screen removed. The sheet is then couched without interposing of felt.

From Korea the art of papermaking progressed to Japan at the beginning of

*Left: Pounding tororo-aoi. Right: Su-keta in action with the help of rods in the ceiling.*

the 7th century A.D. It was a Korean monk called Doncho who, in addition to being well-versed in the paper-maker's art, was a master of various other crafts as well and introduced papermaking during the reign of the Empress Suko. His experiments were none too successful to begin with, however. The paper was brittle and of little use.

It was Prince Shotoku, one of the most eminent characters in the early history of Japan and a great patron of all sciences and arts, who succeeded in improving production methods, partly by using fibres from kozo (*Broussonetia kajinoki*). Thus it was he who laid the foundations of Japan's future greatness as a paper-making nation.

It was not long before the Japanese, as ever, had revised and improved the original Chinese mould and introduced a completely new, specifically Japanese technique: *nagashi-zuki*. Apart from transforming the appearance and function of the mould, the Japanese also invented the method of adding a formation aid, *neri,* to the pulp. Neri thickens the water in the vat, which is an advantage to the paper-maker because the water will then run more slowly through the stock in the mould, giving him more time in which to form a beautiful sheet.

The Japanese paper-maker's mould consists of a hinged frame, *keta.* Fitted to the lower half is a bamboo mat or *su,* which is secured between the two halves by two cleats. The mould is then ready to be dipped in the vat, which is about 75 cm deep and varies in size, that is usually about 1.5 x 2 m.

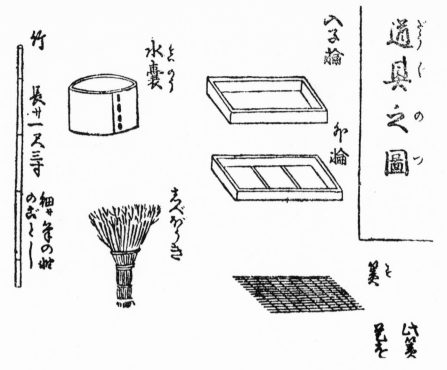

*Japanese implements. Pulp scoop. Brush for laying the sheet against the drying board. The divided keta and su. From Kamisuki Chohoki 1798.*

Because a mould of this kind can be fairly heavy to handle, at least if it is a large one, it is suspended from three resilient bamboo rods under the ceiling which helped to lift it out of the vat.

This kind of mould also has two handles on the upper half. The handles serve two purposes. For one thing they make the mould less unwieldy when working with large formats, and secondly the paper-maker is spared having to plunge his hands into the ice-cold water. In contrast to European hand-made paper production, with the water in the vat kept at about 25°C, the water in a Japanese vat will be as cold as possible. This is because neri looses a great deal of its effect at higher temperatures, which makes it uneconomical to use. A water temperature of about 5°C is not only common but desirable. In the old days, paper-makers also used to have a pail of warm water to one side of the vat to warm their hands in when they got too cold. Paper-makers prefer working in winter, even though the water in the streams they use is ice cold even in summer.

The first step in forming a sheet by the nagashi-zuki method is to put the keta on two support sticks laid across the vat and fasten the su by pushing the strips

*Classical Japanese vatman in action. Note the tub of warm water on the right.*

into the sides, in special grooves made in the keta. Now the su is stretched, two hooks are clipped on to hold the two halves of the keta together, and the mould is ready to be dipped in the stock as soon as the two rods across the vat have been shoved clear.

The mould is dipped vertically into the stock and a suitable quantity is scooped up. Then the mould is tilted forwards, backwards and sideways to distribute the fibres evenly across the su. Surplus water and stock are jolted off. Depending on the thickness of paper required, all one then has to do is repeat the procedure until satisfied.

The keta is now deposited on the support sticks to drain for a while. When sufficient water has run off, the keta is opened and the su is grasped in the left hand by the side furthest away and by the right hand on the side next to the paper-

159

*Taking up stock on the mould.*

*Tossing the stock to and fro over the su, in a wave-like motion.*

*Discarding surplus stock.*

*Couching straight on to the previous sheet.*

maker, whereupon he lifts the su above his head, turning round at the same time, because the paper-maker stands between the vat and the couching board and performs both operations himself.

The first sheet of the day is laid on a felt, just as in *tamezuki,* the western method, but there the similarity ends. Subsequently, to keep all the sheets exactly aligned as they pile up, upright guides are knocked into the couching board. Sheet after sheet can then be couched without any felts in between, because neri prevents the sheets from sticking together, even after pressing.

In couching, the su is deposited on the felt (if it is the first sheet, otherwise on the sheet couched previously). A slight jerk of the strip which the bamboo cloth is sewn on to, and the sheet comes away from the su and settles on the felt or previous sheet.

After couching what one feels to be a satisfactory number of sheets - a post - one may have several hundred piled up, one on top of the other and with nothing in between; the post is then left on its board while the keta and su are cleaned. This done, a cloth is put over the post and another board placed on top of that. The post is left like that until the following day, with just a slight weight - e.g. a bucket containing a few litres of water - on top of the board for downward pressure.

When it is time for pressing the post is carefully transferred to a press, which nowadays will more often than not be a simple jacking device. The Japanese, unlike the Europeans, press their paper with extreme restraint.

After pressing, the original post will have diminished to about one-sixth of its original height, looking for all the world like a rather damp plaster board. Parting and drying one sheet at a time from this compact mass seems utterly impossible until one actually has the opportunity of seeing it done.

A major part in this wizardry is played by tororo, which not only helps to keep the fibres afloat in the vat but also prevents the thin sheets from sticking together and forming one thick piece of cardboard.

Before drying, the sheets have to be parted singly and then dried, either outdoors on a board or else on a preheated drier. Kikuchi uses a vertical, triangular paraffin furnace with enamelled sides.

The sheets are parted by pouring water over their edges and then, when they are slightly moistened, carefully pulling them apart. After that, each damp sheet is brushed onto a drying board or drying furnace. Once dry, they are finished and require no further treatment.

Another striking difference between Japanese and western hand-made paper production is the pattern of movement at the vat. A western paper-maker works mostly with his arms and hands, seldom bending his knees, with the result that his back is subjected to a great deal of strain. A westerner coming to Japan to learn the nagashi-zuki technique finds it difficult at first, before acquiring the more flexible Japanese motion. For my own part I had had a few years' experience of papermaking in Europe, where the technique is based on the vatman scooping stock out of the vat on to his mould, distributing the stock by gently shaking the mould and then, when he has finished, passing it to the coucher. The work posture changes little; the vatman stands virtually still all the time. It takes a long time to switch from tamezuki, with shoulders and arms rigid to the flexed knees, bent back and the rhythmic motions of nagashi-zuki.

My teachers, the paper-making Kikuchi family of Yamagata-machi, drilled me unceasingly in the special movements of nagashi-zuki: "Not like that, Bo-san, you're only using your arms. You must forget tamezuki. Relax, you're too tense. You're working much too hard. If you keep on like that you'll be exhausted before you've made fifty sheets. Try to get into a rhythm. Bend your knees and back. Only your knees must touch the vat. You're much too tense, Bo-san!"

And of course, they were right. Watching my Japanese mentors in action was more like a ballet performance.

There is something spring-loaded about the entire working situation. When the paper-maker pushes his mould down into the vat, the bamboo rods from which the su-keta is suspended help to lift it up again. Perfect balance enables the paper-maker to form his sheet without undue physical exertion.

"You have to let the rods in the ceiling take care of the weight, that's how it works. If you try to lift and force things, you'll only end up killing yourself."

Easier said than done, leaving some of the work to the bamboo rods. My teachers, with a whole lifetime's experience of papermaking behind them, produced one perfect sheet after another with an ease and elegance which they seemed born to. Whistling cheerfully, they manoeuvred the viscous pulp without ever making a mistake. Mould, paper-maker and the rods in the ceiling all seemed part of the same organism.

The mould, su-keta, then, is made up of two separate parts. First a double wooden frame, the two halves of which are joined by copper hinges. And then a bamboo screen which is put on the lower of the two frames and secured by inserting, in grooves along the sides of the wooden frame, the edge sticks to which the bamboo screen is attached.

The keta is built of Japanese cyprus, which apart from being incredibly light is also resistant to water and will not warp. There are only about ten craftsmen in the whole of Japan who make these moulds. Work demands a high level of precision, and every finished keta is a specimen of first-rate artistry and craftsmanship. The light, elegant impression conveyed by the keta is also due to its only being used to hold the su in position during forming. No extra strength is needed in order for the keta to stand up to couching after the European fashion, because the su is removed from the keta before couching. Another advantage of lightness is in the forming of large sheets - 60 x 90 cm is a common Japanese format - when a European mould becomes incredibly heavy and unwieldy.

The bamboo screen, su, consists of two main components: bamboo sticks or splints, and the thread holding them together. The splints begin as bamboo sticks which are drawn through a die to make them perfectly round. The raw material comes from the outer - hardest - layer of wood of the bamboo, preferably bamboo which has grown among cedar trees, because this tends to make it straighter, with more length between nodes. At the splintmaker's house, the bamboo is cut into sections and the nodes discarded. The open tube which is left is cut into pieces and the hardest parts of it drawn through the die. The procedure resembles the Lapp method of producing pewter wire, but whereas the wire is stretched and compressed to the required thickness, in the present case the surplus bamboo is pared away. The usual thickness of bamboo splints in Japan is .625 mm, and they are spaced at about 11 per centimetre. Since, however, the Japanese paper-makers are supreme individualists and all have their own specialities, the variety of sizes, densities etc. for su is practically infinite.

The making of the silk thread is like rope-making on a miniature scale. Each thread consists of three thin strands of silk, spun together in a kind of rig. Just as with the bamboo splints, any number of variations are produced, according to customers' preferences.

Finally the actual su is woven on a wheeled stand, using bobbins. The wheels make it possible for the stand to be pushed to and fro in front of the weaver, so that he will not have to stretch and bend so much. This must be a great advantage, especially when weaving large screens measuring something like 175 cm.

As we have seen, there are many different specialists involved in the Japanese art of paper-making, e.g. su weavers, keta-makers, stick-makers, kozo growers, and suppliers of raw bast. Unfortunately many of these specialities are dying out, and the younger generation are none too interested in such tranquil, unglamorous occupations. In a manner of speaking, these activities remind one of early European paper-making, with specialists at every stage of production, e.g. rag collectors, rag pickers, vat men and bleachers.

Moving closer to European paper-making and to the special kind of paper-making which evolved between 750 and 1000 A.D., let us turn our attention to Samarkand. There is good reason to suppose that the type of mould occurring in Samarkand in about the 750s resembled the current Chinese model, with a mat of grass or bamboo and loose deckle sticks over a rigid wooden frame. As the art of papermaking migrated westwards, however, manufacturing methods had to be altered, due partly to a lack of raw materials to suit the Chinese mould. By the time the art eventually reached Moorish Spain in the 12th century, the appearance of the mould may have changed, but there is no evidence on this point.

Many of the Moorish papers resemble present-day paper from Kashmir, and it is anybody's guess whether the Moors used grass screens. The primitive appearance of the paper may also be due to poorly processed pulp; the raw material consisted mainly of hemp and linen, sometimes mingled with esparto.

Unfortunately we know nothing whatsoever about the type of mould used by the Moslems in either Persia, Egypt or Spain. The same goes for the first types of European mould, with a rigid cloth of metal wire. All we know is that this was invented in Europe or Asia Minor, and the reason must have been that the rigid wire was better suited to the rag pulp used.

In contrast to the oriental method of a flexible grass mat and loose sticks to keep the pulp from flowing over the edges of the mould and back into the vat, this new design had a loose wooden frame which was placed on top of the mould with the wire cloth. This frame, which later came to be known as a deckle, served to limit the size of the sheet and to keep the pulp on the mould while the vatman was working it.

The art of drawing copper wire is closely bound up with the chronology of the first rigid paper moulds. In about 1250 the Italians were producing perfect paper of the very highest quality. By reason of its strength and beauty, the paper produced by Italian paper-makers at that time helped to introduce a new product, and it is to be found today in many European archives.

The same kind of "laid" wire cloth was used, then, as in the Orient, but with copper wire and metal wire substituted, respectively, for the bamboo splints and silk thread. Because the paper sheet is somewhat thinner above the wires than between them, it acquires a faint pattern which can be distinguished when the sheet is held up to the light. Both in the Orient and in the West, the number of "lines" per cm varies a great deal, nor is there any uniformity in the distance between the impressions left by the thinner metal wires linking the copper wires together.

Metal wire was first made by hammering the material into thin plates which were then cut into strips and, finally, rounded with a hammer. The first western moulds must have been made with this kind of wire, because the first known watermark, dated 1282, was formed of metal. Wire-making using the draw-plate came later, and like so many other purportedly European inventions had by that time been used in the Orient for centuries.

This new type of mould, together with improved pulp beaters, did a great deal to improve the quality of early European paper.

*Vat House. From Hallens, 1698.*

The rigid mould consists of a rectangular frame reinforced by cross-ribs, and a detachable frame called the deckle. The earliest European moulds were made of oak, a hard and durable wood. In the south of Europe, use was also made of the softer spruce, pine and cyprus, while Fabriano used poplar. Ash and walnut too were used in France. By and large, use was made of whatever was available.

At the beginning of the 19th century, however, mahogany came to Britain and mould-makers quickly perceived the advantages of this hard, fine porous wood, which is immensely resistant to wet and will not warp even after prolonged use. To make the water run off even faster, the surfaces of the deckle were ground and polished, partly to prevent water from dripping onto the newly formed sheet.

The choice of raw material is critically important. The wood must not warp or split when the mould is being made, and so wood must be chosen which is straight and free from knots. It must be thoroughly dried and, if oak is chosen, it has to be dipped a few times in boiling water, because oak contains a substance which stains the pulp brown. Spruce and pine should also be dipped in boiling water to remove the resins.

Once the wood has been selected, the four pieces of the frame are sawn into shape and carefully planed to absolute flatness. The corners are dovetailed as accurately as possible, to prevent water and pulp penetrating the joints.

*Vat house. From de la Lande, 1762.*

The wood for the four parts of the frame should be about 8-10 mm thick and 35-40 mm high. In certain cases - e.g. extra large moulds - the frame can be up to 60 mm high. To protect the mould from wear when it is being pushed from the vatman to the coucher, the frame is usually fitted underneath with a reinforcement of brass or hard wood.

When the corners have been dovetailed and everything fits together as it should, it now remains to make the ribs supporting the metal wire which will subsequently be fitted on top of the frame. These ribs are roughly the same thickness as the frame, but the upper part of each one is planed or cut to an edge, so that the water will run off more easily instead of collecting on the top. There is also a peg hole at each end. The ribs are placed at intervals of about 20 mm and the positions of the pegs are marked, after which holes are drilled accordingly along the sides of the frame.

Before the ribs and frame are assembled, a number of small holes are drilled in the tops of the ribs. The metal wire securing the metal cloth to the ribs will be drawn through these holes. The frame and ribs are then glued together with water-resistant glue.

167

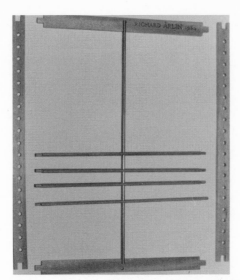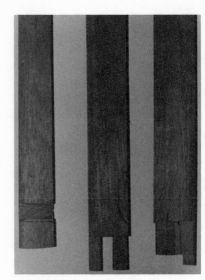

*Left: Structure of the mould, with ribs and brass rod. Right: Detail of a dove-jointed deckle.*

As stated earlier, all paper moulds of European origin have a deckle. As a matter of interest, when the Tumba Mill produced its three-ply bank note paper, the deckle was only used for the centre ply, i.e. that carrying the watermark. The thin top and bottom plies were formed without a deckle - a task which must have required a great deal of skill on the part of the vatman. Usually at least two deckles were made for every mould, as it is subjected to quite considerable wear, due to being constantly applied and removed by the vatman.

The deckle serves to retain the appropriate quantity of pulp on the mould until sufficient water has drained off. At the same time, the inside of the deckle determines the size of the sheet left on the mould. Depending on how well it fits, it will "cut" the wet sheet in different ways. A perfect fit means straight sides and very perceptible deckle edges. A "leaky" deckle means distinct, uneven deckle edges - an aspect of hand-made paper which many people find charming and an indication of its being genuinely hand-made.

The deckle covers the whole of the top of the frame and about two-thirds of the outside, as well as roughly 10 mm of the wire. Making the deckle is an even more skilled task than making the frame. The frame, with its ribs and dovetailed corners, is rigid in itself, but the deckle only derives its rigidity from the complicated dovetailing of the corners.

Precision in dovetailing the corners has a critical effect on the ability of the deckle to resist the strains imposed by forming - between 1,500 and 2,500 cycles a day in professional use. The deckle has to stand up to this treatment for years and at the same time remain exactly rectangular.

The type of dovetail joint used in the classical European deckle is a combination of dovetail and ordinary recessing. But it is so ingeniously done that it must be termed a first-rate specimen of precision carpentry. Oddly, enough, this type of dovetailing has been found in old deckles from all over Europe and also in the USA. Perfectly executed, it guarantees the rigidity of the deckle over the years. A few deckles have been found with ordinary mitred corners, but this is a sign of poor craftsmanship more than anything else. A mitred joint can possibly be made secure with the immensely strong glues available nowadays, but only a corner joint of one kind or another was good enough for the old European mould-makers.

When the corner joints have been cut to shape and the deckle is exactly rectangular at the corners, the wood is planed down about 10 mm, but still leaving an edge about 10 mm wide. In the finished deckle, the inner dimensions of this edge will correspond to the external dimensions of the frame.

Next a "channel" about 25 mm wide is carved between the outer edge, corresponding to the outer edges of the frame, and the inner edge of the deckle, which will rest against the metal wire cloth and limit the size of the sheet. The long sides of the deckle are then ground down until they slope down towards the wire cloth. This will enable the vatman to discard surplus stock. The ends are also bevelled down towards the wire, and four thumb holds are ground into the wood, so that the vatman will be able to grasp the deckle firmly.

To make the deckle "cut" better, i.e. lie as flush as possible with the frame and wire cloth underneath, thus preventing the pulp from running out, its sides can be made slightly concave and its ends slightly convex. Then, when the vatman depresses the deckle to grasp the ends, the sides will butt against the frame and wire underneath.

When the four parts of the deckle fit together exactly, the joints are glued with water-resistant glue.

It now remains to make the wire cloth and attach it to the frame. As we have already seen, the rigid mould and its permanently attached wire cloth presumably originated some time during the 12th century, somewhere between Asia Minor and Italy. Since it was at this time that the Europeans learned the art of drawing wire, the first rigid moulds can be more or less reliably dated. Copper, presumably, was the raw material used.

The rapid growth in demand for paper for records, teaching and commerce led to a snowballing of papermills. All of them used moulds, most of them fairly short-lived, but nearly all of them with a mark denoting the feudal lord, the owner of the mill or the client's monogram.

Although not all moulds had a watermark, all of them did have a wire cloth made of copper wire. Making this cloth was a time-consuming business, and so

*Frame waiting for its covering.*

*Next page: Richard Årlin's technical drawing of a classical European paper mould.*

it is tempting to assume that the thought soon occurred of mechanising the procedure. Eventually this resulted in the invention of a loom for wire-weaving. We do not know when or where this invention was made, nor by whom. The earliest mention of a wire-weaver's loom occurs in a petition to the English Privy Council in July 1694, when Nicolas Dupin states that he is able to manufacture moulds much faster than anybody else. The earliest Swedish record of such a device dates from 1788 according to Ambrosiani *(Dokument rörande äldre pappersbruken i Sverige, p. 194).*

There are two versions of this wire-weaver's loom: horizontal and vertical. The horizontal version is basically similar to a large textile loom, while the vertical one consists of standing wooden loom fitted with hooks at the bottom. These hooks secure the thin metal wires constituting the warp. The upper part of the loom has a set of fasteners which are twisted to secure the other end of the warp.

The procedure for making the metal wire, quite simply, is to insert bronze wire in the warp, turn the upper fastenings through 180° so that the warp will hold the wire in position, and then start all over again.

Several such frames are extant in Sweden today, e.g. at the Tumba Bruk Museum, Ösjöfors, Lessebo and the Göteborg Industrial Museum.

The Tumba loom used an endless wire which was kept permanently straight and stretched by two screws and was passed through the warp with a shuttle. At the end of each pass, the shuttle would turn upside down, whereupon the wire fell out and landed in position on top of the previous layer. Once in position it was cut off and knocked home by the beam, and the rotating fasteners for the

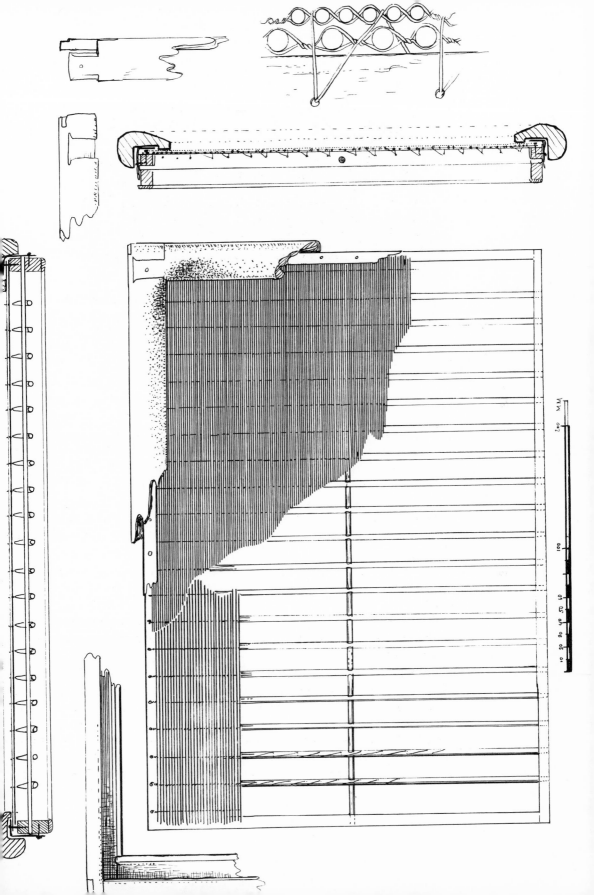

*Putting the wire in position between stainless warp threads.*

thin metal wires of the warp were turned through 180°. A new wire was inserted in the shuttle and passed through the warp, and so it went on until the wire cloth attained the necessary size.

Needless to say, the tension of the warp thread had to be adjusted as work proceeded. When the cloth was finished, the warp thread was twisted for a length of some centimetres, and this part of the wire was used for securing the cloth to the frame.

A wire cloth of this kind is termed "laid", in contrast to the genuinely woven metal cloth used for making a paper like that ordered by the famous English printer John Baskerville from the Balston Paper Mill in Maidstone, Kent, England. That mill was run by James Whatman, and tradition has it that the loom in which the wire for Whatman's mills was produced was designed by the mould-maker *John Sellers* of Chepstow.

A loom of this kind must have been of a completely different design from that used in Tumba. In his book *"Paper Mould and Mould Maker"*, Edo Loeber writes that the early looms for "woven" metal cloth needed three workers: one on the left, one on the right and one in the middle. Those on the left and right passed a shuttle, loaded with copper wire, from side to side, while the man in the middle operated the long beam and reed extending the full width of the frame. Every wire was beaten into position, an immensely fatiguing task which for the

*Wire-weaver's loom from Lessebo Bruk. Notice the shuttle.*

most part was left to the apprentice. The right-hand man attended to the lifting of the warp, so that the shuttle would be passed alternately between lowered and raised warp wires. Basically, this technique corresponded to ordinary tabby weave, while the Tumba frame was more like a kind of insertion weave.

The reason given by John Baskerville for his order was that he wanted a paper of softer appearance and without any laid lines. True enough, the woven cloth did not leave so many traces in the finished sheet. The impression was more like that of a fabric. The first book to be printed on "wove" paper was Baskerville's magnificent edition of *Publii Virgilii Maronis Bucolica, Georgica, et Aeneis*, published in Birmingham in 1757.

173

*Horizontal wire-weaver's loom from Amies & Sons, Maidstone.*

Baskerville's desire for the least possible trace of the wire could not be completely accommodated. The mould-maker had evidently sewn the wire straight on to the ribs, with the result that the paper acquired a striped pattern. In about 1800 this "defect" was overcome by putting a coarser metal screen underneath the wire cloth.

Baskerville, however, did not always insist on wove paper or *papier vélin,* as it came to be called. His very famous 1769 Bible, for example, was printed on antique laid paper.

That which Baskerville was at pains to eliminate from his paper, we today regard as original and beautiful. Machine-made, antique laid paper is manufactured using a dandy roll to produce the pattern which used to be left by the laid wire. The thin metal wire of the warp was twisted round the thicker weft wire, producing a chain-like impression in the pulp. This runs across the paper, while the laid lines were made by the thicker wire.

In a modern European hand-made paper mould, the wire covering consists of two layers: a bottom layer of about 1 mm brass or German silver wire, and another one of about 0.7 mm bronze wire. The upper layer is denser than the lower.

The lower, coarser covering is attached to the sides of the frame with brass or copper tacks (materials that are capable of rusting must never be used in

*Left: Forming the sheet. Right: Removing the deckle.*

*Left: The coucher receives the mould. Right: Putting the mould to drain.*

*Left: Couching the sheet. Right: Putting a felt on the new sheet.*     175

*Woven wire covering of phosphor bronze wire with watermarks sewn on.*

papermaking, because even the slightest flake from a steel wire or suchlike will eventually cause brown stains - "foxing" - in the paper).The wire covering is then sewn on to the ribs with stainless metal wire.

The upper, finer covering is then sewn on to the lower one. Any watermarks are sewn on to the upper covering.

Once the wire coverings are in position, the mould is completed with a thin brass plate which covers the upper part of the frame and extends about two-thirds of the way down the sides. Over the centuries, the sizes of European moulds have varied to such a degree that practically every conceivable size has been used which could be held between two arms. The very commonest sizes, however, are in the region of 30 x 40 cm paper size, i.e. the area delimited by the inside edge of the deckle.

The battle of paper against parchment also resulted in the size of the new material being adapted to that of the old. In many cases the two were kept in the same standardised boxes and they were even bound together in books.

When making up their orders, the mould-makers also had to bargain for the tendency of paper to shrink when drying. The paper-maker in turn was more interested in keeping the number of sizes to a minimum.

For smaller formats he could insert a dividing rib and halve the sheet size, or else he could attach ribs to the inside edge of the deckle and reduce the size in this way.

For envelope-making, special deckles were constructed which limited the area of the wet sheet to the shape of the envelope. When, during the 1840s, envelopes came to be more frequently used, superseding the old method of folding a letter up and sealing it at the edges, this presented paper-makers with a new lucrative market. The innumerable paper formats demanded just about the same number

*Left: Attaching the lower, coarser wire covering to the frame. Right: Sewing the wire covering on to the ribs.*

of envelope formats, and there was never any question of standardisation. There were two ways of manufacturing envelopes. Ordinary rectangular sheets could be produced and then cut to shape, or else the mould could be fitted with cutting wires the shape of an envelope. The Göteborg Industrial Museum has a mould specially constructed for envelope paper and restricting the area of the wet sheets to the exact shape requirements.

Filter paper was another speciality of the hand-made paper-makers. In the contest with machine-made paper during the mid-19th century, production of circular papers was one of the last "niches". Mostly a deckle was used which, with the aid of covering plates, left a circular space in the middle, but a complete circular mould and deckle have been found in the German hand-made papermill at Hahnemühle.

Visiting cards were another article requiring special moulds. The thicker grade of paper called for a higher deckle, and the wire covering was fitted with some sort of turn-off device. Mostly this took the form of sewn-on coarse copper wire or else a soldered metal profile. After couching, this left a wet sheet with a squared pattern, each square corresponding to a visiting card. After pressing and drying the full sheet would be torn to a smaller format, along the lines left by the wires.

*Envelope mould from Holmens Bruk, Motala, c. 1890. Photograph: Göteborg Historical Museum.*

The rather chaotic product range of the hand-made paper-makers - after all, they had to suit all tastes - was gradually slimmed down as improvements were made to papermaking machinery. The machine age also changed people's attitudes to many every day articles. The fact was that things could be made well in machines, and ever since Klippan installed its own and Sweden's first paper machine in 1832, consumption has risen uninterruptedly. Hand-made paper is a minute, barely quantifiable proportion of present-day output, but it is important, both historically and culturally speaking.

# THE PAPER MACHINE

### From small to large-scale industri

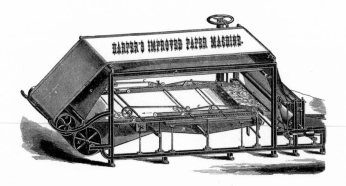

*Previous page: Harper's improved paper machine, c. 1860. From Hofmann.*

Machine papermaking is based on the same principle as hand-made paper production. The work of the vatman, when he forms his sheet by manipulating the mould - his "stroke" - is done in the machine by transferring the pulp to a metal wire on which it is shaken and de-watered. It is then taken to be pressed and dried and perhaps finished in other ways, e.g. by glazing. This latter work, though, is often done in special machines and is a separate finishing operation. Sizing is done already at the pulp stage.

The early papermaking machines were of two main types: the fourdrinier and the cylinder. The former, invented in 1799 by Louis Robert, but called after the Fourdrinier brothers, who greatly improved the basic model, consists essentially of an endless wire cloth stretched between rolls. The wire forms a flat horizontal or slightly inclined surface onto which the pulp is poured.

The cylinder machine has a wire screen fitted like a jacket round a cylinder, which is immersed in a vat of pulp. This type of machine is mainly used for producing cardboard and other coarse grades, e.g. duplex or triplex paper, that is paper consisting of two or three layers of pulp worked together in the couching press.

Sweden's first papermaking machine, installed at the Klippan Mill in 1832, was a Fourdrinier machine with a total length of 12.5 metres. The working width was 1.35 metres and the wire from breast roll to couch not quite 4 metres. Felt was only used in the first press; in the second, the web of paper passed between steel rolls, which can account for the surface of Klippan machine-made paper being so outstandingly smooth and even compared with the hand-made variety. This was a precursor of the glazing press which later became so common.

After the wet press came the drying apparatus, consisting of five drying cylinders with a diameter of 0.7 m. The three lower cylinders had dryer felts, while the two upper ones had none. The calender was a "fourth press", installed downstream of the dryer section. Finally the web was reeled up on spools which, to simplify cutting operations, were mounted in a movable frame. Steam from a boiler was carried through iron and lead pipes to heat the dryer cylinders.

With new inventions and improvements occurring in rapid succession, paper machines were frequently altered and enlarged. Towards the end of the 19th century, developments began to move faster than ever, although machine manufacturers at that time were still not able to introduce all the improvements that paper-makers wanted. Gradually, though, various changes were introduced

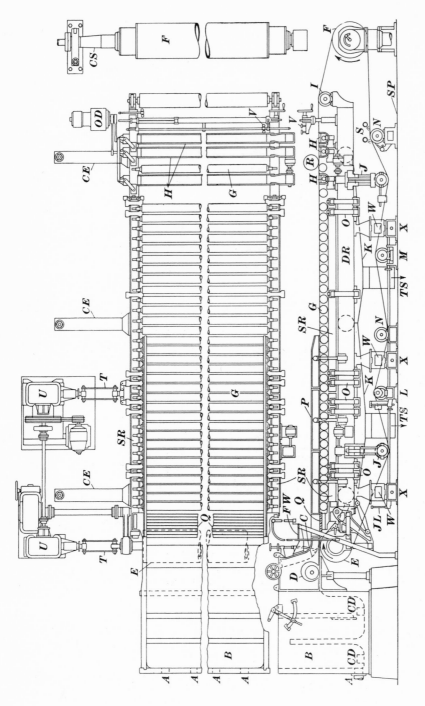

*Conventional Fourdrinier machine. A, stock inlets. B, headbox. C, slice. D, distributor roll. E, breast roll. F, couch roll. G, table rolls. H, suction boxes. I, turning (end) roll. J, stretcher rolls. K, save-all pans. L, automatic guide roll. M, hand guide roll. N, wire rolls. O, supports. P, deckle. Q, forming boards. R, dandy roll. S, showers. T, shake connectors. U, shake mechanisms. V, cut squirt. W, rise block. X, jacks. CD headbox washouts. CE, cantilever extensions. CS, couch shaft. DR, "dead rail". FW, foot walk. JL, jack lever. OD, oscillator drive. SP, sole plate. SR, shake rail. TS, save-all pan drains. From Stephenson, Pulp and Paper Manufacture.*

which elevated the quality of paper machinery. The working width was increased but still did not exceed 3 metres. The wire, press and dryer sections were enlarged for faster through-put.

Demand for paper, which was escalating all the time, was a severe test for the machine manufacturers. New paper qualities were in demand for any number of purposes and, apart from the technicalities of papermaking as such, new problems arose because, very often, the new qualities called for modifications in the design of the paper machinery.

Advances in printing technology enabled printers to work with larger formats and with much faster machines. Rotary printing was introduced by the main newspaper companies and a succession of new methods such as off-set, photogravure and collotype now saw the light of day.

Fast paper machines, especially for newsprint, already existed on the drawing board by the turn of the century. Actual production speeds, though, seldom exceeded 100 m. per minute, which by the standards of the time was fast indeed. Fine paper production, however, had to make do with speeds of about 50 m/minute.

Things moved faster where lower qualities were concerned, and already by about 1915 widths had reached 5 metres and working speeds about 200 m/min. This record was soon broken, for by 1920 a machine with a working width of 5.8 metres and a speed of 300 m/min had been designed in the USA. Ten years later the Great Lakes Paper Co., Fort William, Ontario installed a printing-paper machine with a working width of 7.7 metres and a speed of 395 m/min.

By the mid-1930s, the working width was 8 metres and the speed 400 m/min, but for all these improvements in efficiency, the basic design was the same as ever: a wire section, a press section and a drying section.

The wire section consisted of an endless metal cloth stretched between two thick rolls. One of these, the breast roll, was positioned at the point where the stock was poured onto the wire, and the other, the couch roll, was next to the press section. Next to the breast roll the wire was supported by calender rolls which, in addition to supporting the wire, also helped to drain off the water from the web.

In the earliest machines, the distance between the calender rolls was shortest next to the breast roll, but in faster machines the opposite applies, so that the paper web can be formed more smoothly.

For optimum paper formation, stock has to be suitably diluted before flowing out onto the wire. It is equally necessary to have the right drainage. When the stock is poured out onto the wire, it is also subjected to a short, sideways shaking movement, the mechanical counterpart to the vatman's stroke.

Because the pulp is so heavily diluted when it enters the wire, some device is

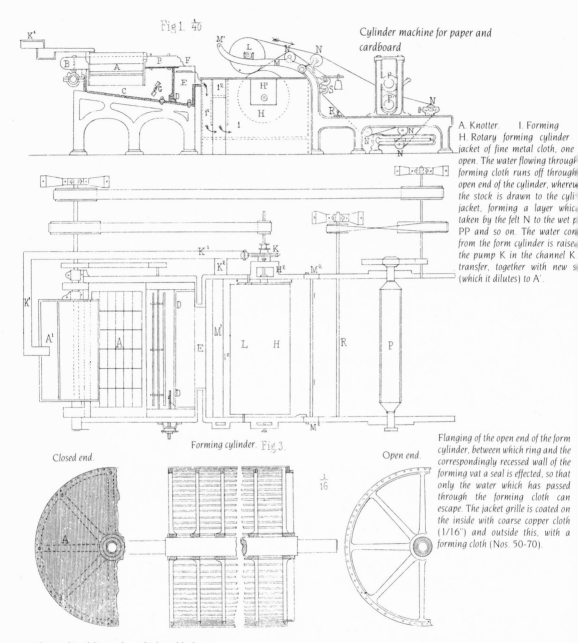

Fig 1. $\frac{1}{40}$

Cylinder machine for paper and cardboard

A. Knotter.     I. Forming
H. Rotary forming cylinder
jacket of fine metal cloth, one
open. The water flowing through
forming cloth runs off through
open end of the cylinder, whereu...
the stock is drawn to the cyli...
jacket, forming a layer whic...
taken by the felt N to the wet p...
PP and so on. The water co...
from the form cylinder is raise...
the pump K in the channel K...
transfer, together with new s...
(which it dilutes) to A'.

Forming cylinder. Fig 3.

Closed end.

Open end.

$\frac{1}{16}$

Flanging of the open end of the form
cylinder, between which ring and the
correspondingly recessed wall of the
forming vat a seal is effected, so that
only the water which has passed
through the forming cloth can
escape. The jacket grille is coated on
the inside with coarse copper cloth
(1/16") and outside this, with a
forming cloth (Nos. 50-70).

This machine delivers a loose, fairly unfelted paper.

Paper machine parts. Pulp vat and agitator.
The agitator should work in such a way that the fibres will not be lumped together by eddies.

Regulator pump for dosage of stock on a shaking paper machine.

Fig.1.

Raked paddles.

45 à 25 mm.

3 à 4 meter

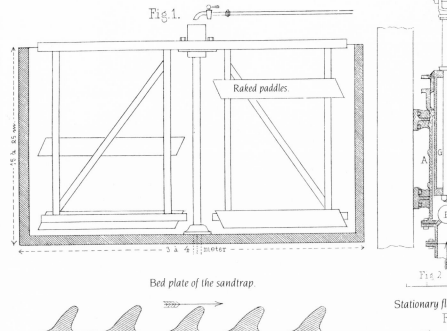

Fig.2.

F

G

A    G    C

B

The length of the stroke is altered to match the quantity of pulp to be fed onto the paper machine. The pump should be powered by the same motor as the machine.

Fig 2.   Scale.          50 cm.

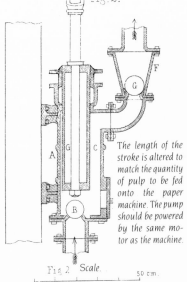

Bed plate of the sandtrap.

Stationary flat knotter.
Fig 6.

The stock is shaken in the screen openings by the reservoir bed (a) (a stiffened rubber plate) moving upwards and downwards.

Cross-section of a strainer plate.
Fig 5.

Flat double shaking screen.
Fig. 4.
Cleaning position.

a    b
c

a. First screen   b. Check screen

Fig. 6.

a

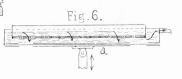

Rotary knotter with suction and pressure piston.
Fig. 7.

150 T   P   6T

k

R   lyrkintig   silprisma

L

K

The pulp enters from the reservoir K through the square, rotating screen prism R and flows out through the hollow spigot. The pulp is agitated in the screen openings by a reciprocating piston in cylinder P. The screen plates in the knotters are made of metal and are fitted, in a special machine, with screen slits (Fig. 5). The width of the screen slits is adjusted to the type of paper, with at least 3-4 sq.m. screen area for an ordinary paper machine. A finer check screen is of advantage.

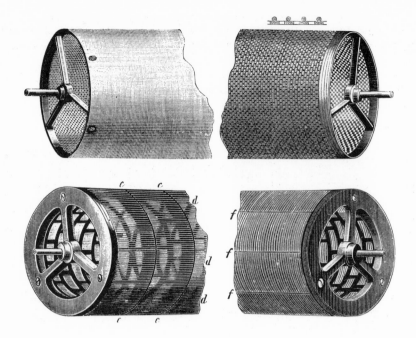

*Dandy roll. From Hofmann.*

needed to prevent it spilling over the edges. The earliest device of this kind consisted of deckle straps running parallel to the wire and keeping the pulp in position.

After the tube rolls, the wire passes over a number of suction boxes. These are boxes, 20-30 cm wide, of stainless material which straddled the machine and "vacuum-clean" the water from the stock.

A dandy roll is positioned over the web to make the paper more even and to give it a more attractive transparency. This light roll, about 30-80 cm in diameter, rests under its own weight on the web and is impelled by the latter. By attaching figures or letters of the alphabet to the metal netting on the dandy roll one can leave an impression in the stock, and in this way produce a watermark.

The turning point of the wire is occupied by the couch roll, where the paper web is further de-watered. The American Millspaugh constructed a "suction couch" as early as 1909, by perforating one of the couch rolls and connecting it to a powerful vacuum pump.

By the time the paper has passed through the suction couch, it is generally so dry that it can support itself all the way to the press section. This consists of a number of pairs of rolls - wet presses. These rolls are felt-covered and they remove as much water from the web as the greasiness of the pulp will allow. As

*Yankee machine, 1950. From Althin 1.*

the web advances through the rolls, its dry content rises to between 30 and 40 per cent.

After pressing the web comes to the dryer section. This consists of a number of hollow, steam-heated drying cylinders, usually ranged in two horizontal lines, one above the other, with the upper line displaced in such a way that each of its cylinders comes midway between two of the lower ones. Progress through the drying section is greatly simplified by the web alternating between the upper and lower cylinders.

After the drying section there are usually one or two cooling cylinders to give the paper a suitable, consistent moisture content. 50 per cent relative humidity is desirable for further processing of the paper.

One very common type of machine in Scandinavia is the "Yankee machine", the foremost characteristic of which is that it has only one drying cylinder, the Yankee cylinder, which on the other hand has a very big diameter. The Yankee machine is mostly used for making thin and medium-thin paper for wrapping

189

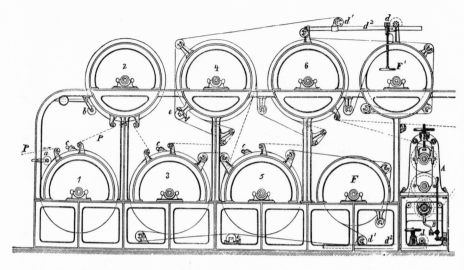

*Drying plant. The web, P, passes through the drying cylinders 1 and 2 and from there to the cylinders with drying felts 3, 4, 5, 6 through the damping press A and from there to cylinders 7, 8. Cylinders F and F2 act as felt dryers. The worm-gear (D) actuates the stretch roll (D1) via the pinion gear (D2). From cylinder 8 the web continues past the brush damper D to the calender E and the slitter H, proceeding from there to the rewinder J. From Hofmann.*

purposes, known as MG paper. MG is short for "machine glazed", referring to the high gloss of the side of the paper butting against the drying cylinder.

The Yankee machine is often equipped with web transfer, which means that a woollen felt, the pickup felt, has been arranged so as to encircle the couch roll. The web is then pressed against the pickup felt and adheres to its underside. Together with a bottom press felt, the pickup felt and the paper sheet adhering to it then pass through a wet press, also known as the bottom press.

From this wet press the web, together with the pickup felt, passes through a felt guide roll into a cylinder press which presses the web under heavy pressure against the Yankee cylinder. In this way the web is transferred to the surface of the Yankee cylinder.

The good contact between the paper and the hot surface, combined with good ventilation and pre-heated dry air and high steam pressure in the Yankee cylinder, makes for a very high rate of evaporation per sq. m. drying surface (about five times more than in a multicylinder machine).

A kind of combined paper machine, used specially for producing duplex cardboard, is fitted with a pre-dryer section consisting of between 3 or 4 and 18 or 20 drying cylinders. After this section comes the Yankee cylinder, and then there are a number of after-drying cylinders for the final drying of the paper.

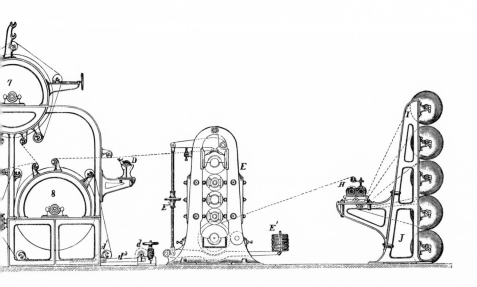

In the cylinder machine, otherwise known as the vat machine, the web is formed round a rotating cylinder also known as the cylinder-vat unit. The cylindrical surface of the cylinder-vat unit is faced with a fine metal cloth and immersed in a pulp trough to which the stock is supplied in highly diluted form (0.08-1 per cent dry content).

The web is formed by keeping the level of liquid inside the rotating cylinder below the level of the stock on the outside. In this way the water in the stock is pressed through the metal cloth and the pulp fibres are pressed by the pressure of the water against the metal cloth. In this way a layer of paper is built up which is then couched onto an endless felt - the pickup felt. The vat machine has been used almost solely for producing relatively thick cardboard, and so several cylinder-vat units were built in line, so that the pickup felt would pick up the layers of paper one after another and couch them together. As has already been made clear, paper machines have changed a great deal in both speed and size during the 150 years or so for which they have now existed. Sweden's first paper machine, the PM1 at Klippan, was 12.5 metres long when installed in 1832. Today's computerised giants are of completely different proportions.

The KM7 at Skoghallsverken is 240 m. long with a wire breadth of 6 metres. It has a maximum speed of 450 m/min and an annual output capacity of more

191

than 200,000 tons. The drying equipment consists altogether of 83 drying cylinders, viz 47 upstream of the MG cylinder, 32 cylinders in the after-drying section (16 upstream of the sizing press) and 4 in a small swing dryer downstream of the coating section.

The MG cylinder (the Yankee) is six metres in diameter. The KM7 being purely a cardboard machine for the 200-500 $g/m^2$ grammage range, the main task of the MG cylinder is to minimise the surface roughness of the cardboard - an essential prerequisite of good printing.

# THE CHEMISTRY OF PAPER

Arnulf Hongslo on the structure of the material

*Previous page: Early wood pulp boiler.*

All the paper manufactured until 1845 was textile fibre paper or rag paper, as it is often called, but in the mid-1840s came the invention of ground pulp, made by grinding wood. Whereas rag pulp contains fibre of pure cellulose, ground pulp consists of wood fibre fragments containing about 50 per cent cellulose fibre and the rest lignin, which serves to bind the cellulose fibres of the wood, together with a small amount of hemicellulose and polysaccharides. The lignin has no fibrous structure but its chemical composition is highly complicated and it has other properties which impair the quality of paper. For example, the paper turns yellow very quickly and its mechanical strength declines, causing it to become brittle and to disintegrate after a time. Nor can paper be manufactured exclusively from ground pulp; usually 15-20 per cent ordinary chemical pulp is added, to achieve sufficient strength.

For a few decades following the introduction of mechanical wood pulp in papermaking, it was common practice for paper to be made from ground pulp mixed with rags. The disadvantages of ground pulp had not yet been realised, and since there was a shortage of rags, ground pulp was used for practically all paper. Before long, though, it was discovered that this paper was not equal to prolonged storage. This was particularly true of archival paper, i.e. paper intended for documents, survey maps, legal records and so on. In time, archives had so much trouble with the storage of these papers, because they turned yellow and fell to pieces, that legislation had to be passed defining the make-up of the paper used for these purposes. The first Archival Paper Ordinance was promulgated in 1907. In the early 1960s it was revised and new provisions added.

The most important of these provisions is that archival paper must be made from 100 per cent textile fibre (cotton, flax and hemp). Paper made from these fibres, properly manufactured, must keep indefinitely, without yellowing to any appreciable extent and without any impairment of its strength properties.

Archive-quality paper initially contains fibres with long cellulose molecules, long chains of glucose molecules, which accounts for the high strength of the fibres. In an acid paper, i.e. a paper sized at a low pH value, these long molecular chains are rapidly broken down, especially if the paper is stored at high temperature or in a sulphurous atmosphere. This weakens the individual fibres, thus impairing the strength properties of the paper.

To counteract the acidity of archival paper, it is common practice nowadays to use neutral sizing, i.e. to size the paper at a pH value close to the neutral point.

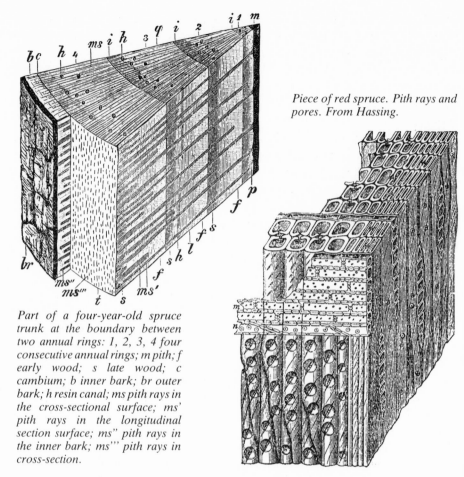

*Piece of red spruce. Pith rays and pores. From Hassing.*

*Part of a four-year-old spruce trunk at the boundary between two annual rings: 1, 2, 3, 4 four consecutive annual rings; m pith; f early wood; s late wood; c cambium; b inner bark; br outer bark; h resin canal; ms pith rays in the cross-sectional surface; ms' pith rays in the longitudinal section surface; ms" pith rays in the inner bark; ms''' pith rays in cross-section.*

In addition, filler is added containing calcium carbonate (chalk), which neutralises any acid in the paper and prevents acidity from increasing.

To ensure that the paper really has the necessary strength properties to begin with, there are official regulations concerning tear strength and weight standards. There are also regulations concerning the chemical properties of archival paper. Acidity rates and pH values are very important here. The quality of the raw material is often checked by working out what is known as the copper number, indicating the quantity of pulp degradation products (oxy- and hydroxy-celluloses). Nowadays, however, it is very common to carry out viscosity determinations in order to determine the extent to which the cellulose material has been degraded.

The modern paper industry would never have become what it is today if a method had not been found of releasing the cellulose fibres in wood. The procedures used today were developed during the second half of the 19th century, and they are based on cooking with suitable chemicals to dissolve the lignin and remove it from the cellulose.

The original soda pulp method was developed into the sulphate pulp method, which is still in use. With this method, the wood chips are digested in pressure cookers with sodium sulphide (Na2S) and caustic soda solution (NaOH). This method has been used in Sweden since the mid-1880s. Pulp made by this method is brown and very strong, and it is mainly used for wrapping paper. It was not until 1930 that a way was found of bleaching this pulp so that it could be used for white paper. White pulp, however, was already made in the 1870s by the sulphite pulp method. In this case, the wood chips were digested by pressure-cooking them with calcium bisulphite, which effectively dissolved the lignin.

Nowadays sulphite pulp is also cooked with magnesium bisulphite. This "magnefite" pulp is produced at a pH value of 4.5 and is easy to bleach, nor is it so easily degradable as ordinary calcium bisulphite pulp. The procedure is used when cooking pulp for fine paper, but not so often as the sulphate procedure.

As mentioned earlier, it was quite a long time before effective bleaching methods were developed for sulphate pulp. The original hypochloride bleaching method, which had been known since the end of the 18th century, was good enough for sulphite pulp but not for sulphate pulp. It was not until methods like chlorine dioxide bleaching and hydrogen peroxide bleaching were developed that bleaching of pine sulphate pulp became possible to bleach. The bleaching methods now available yield not only whiter pulp but also pulp with better properties in other respects, e.g. superior strength and longer durability. This is due to a more selective bleaching, whereby the bleaching agent breaks down the colouring substances in the pulp without noticeably affecting the cellulose. Where modern rag preparation is concerned, it is worth adding that the earlier method, which involved boiling with ordinary soda or caustic soda and then a separate bleaching operation using hypochloride, has been superseded by a method used in the textile industry, viz pressure boiling with hydrogen peroxide in an alkaline environment. This method gives a half-stuff of superior whiteness and strength compared with the old method.

Bleached und unbleached pulps are used in large quantities in the paper industry. In addition to mechanical (ground) pulp, use is also made of a semi-chemical pulp in which only part of the lignin has been removed. Unbleached pulps are extensively used in the packaging paper industry, and a great deal of ground pulp is used for newsprint. Rag paper used to be made from textile waste and old cotton and linen. Nowadays mostly cotton linters from the textile industry are used instead.

Fibrous materials like straw pulp and esparto pulp (made from esparto grass)

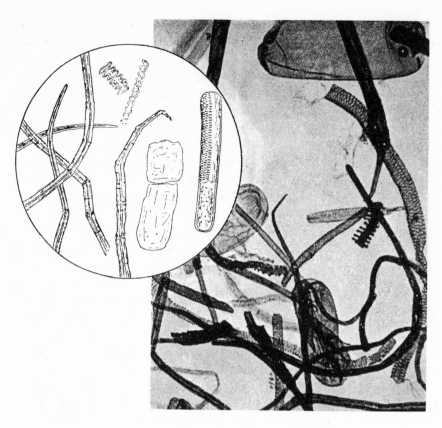

*Straw cellulose, magnified 250 times. From Hassing.*

are very seldom used in Sweden nowadays. They have been replaced by softwood pulp (birch and other broadleaf woods). The short fibres of broadleaf trees do not make strong paper, but they make a good additive to stock for offset paper. They then act as a kind of filler, helping to give the paper a smooth, even surface.

It would be going too far in this connection to describe the structures of all different kinds of paper. Composition naturally depends on the use for which the paper is intended. Bleached sulphite and sulphate paper, for example, occurs in most kinds of printing and writing paper. The same goes for map paper and drawing paper, but in these cases the chemical pulps can also be mixed with rag pulp (textile fibre). Certain kinds of art paper, e.g. watercolour paper, lithograph paper and copper-engraving paper, also contain a certain quantity of rags. The aim here is to produce a paper which will not turn yellow too much when exposed to light. Paper for copper engravings also has to have a high bulk, which facilitates the absorption of ink from the plates.

To understand the structure of a paper and the forces holding together its constituent fibres, one needs to know the fine structure of the cellulose fibres. Only then can one appreciate the immense importance of the grinding of the fibrous pulp as a factor influencing the strength and other properties of the paper.

Cellulose is a chain molecule. Every link in the chain is a glucose molecule, i.e. a sugar species with five carbon atoms and one oxygen atom. A sixth carbon atom is attached to this ring, and this in turn carries a hydroxyl group (-OH group). Every glucose molecule has three hydroxyl groups; the other two are linked directly to the carbon atoms of the ring.

The glucose molecules, or the rings, are held together chemically with one oxygen atom (O) between each pair of rings. The cellulose chains can vary in length; sometimes several thousand can be joined together, one after the other. If so, this is a very strong cellulose (alpha cellulose). Some chains, however, are a good deal shorter. These chains form part of what is called hemicellulose, which is really something of a hybrid between cellulose and sugar species.

The chain molecules are positioned side by side in the cellulose substance, forming, as it were, layers stacked on top of each other into packets. These packets are sometimes neatly arranged, but at the sides and in between the packets the chains are more or less haphazardly structured. The parallel cellulose chains form bundles - fibrils - which make up the actual fibre.

The fibre is rather lika a hollow cable, with a hole in the centre - the lumen - and a wall consisting of several concentric layers of fibrils. Textile fibre - cotton, for example - is surrounded by a thin casing consisting of cellulose fibrils, pectrine and wax substances. This casing is called the primary wall. Beneath it is the thicker secondary wall, consisting of several concentric layers of fibrils, arranged in spirals which slope slightly along the length of the fibres. Wood fibre has another thin layer outside the primary layer, but this disappears during digestion.

The thickness of the fibrils and the angle of the different fibril layers to each other vary from one kind of fibre to another.

In the case of linen fibre, the fibrils are fairly coarse and perfectly distinguishable in a microscope. The fibrils in cotton fibre are not quite so coarse but still visible through the microscope. In wood fibre, the fibrils are a good deal finer than those of flax and cotton. These "sub-microscopic" fibrils are not visible through an ordinary optical microscope, nor do they occur in such large quantities as in flax and cotton fibres.

The big difference in thickness, quantity and structure between textile fibre and wood fibre considerably affects the grinding procedure and the strength development of the pulp. Textile fibre, with its relatively large quantities of coarse fibrils, gives a much stronger paper than wood cellulose fibre.

Whereas textile fibres like cotton and linters consist mainly of pure cellulose, wood cellulose fibres contain much more hemicellulose and sugar species. As regards size, textile fibres can be very long, anything up to 25-30 cm. Wood cellulose fibres, on the other hand, are far shorter. Conifer cellulose fibres, for example, are about 3 mm long on average, while broadleaf wood cellulose fibres are about 1 mm. In other ways too, fibres vary quite considerably in shape. Broadleaf fibres, for example, are much wider than conifer fibres, and they also have a more complicated structure. Then again, the shape of wood fibres can vary quite considerably in one and the same tree, the reason being that, while they were still cells in the living tree, the fibres had different functions.

Basically, a sheet of paper is a network of fibres. It is made by first dispersing the fibres in water and macerating them either in hollanders or in refiners. The pulp thus produced is further diluted with water and pumped to a wire which moves forward at a certain speed. The water now runs straight through the wire, while the fibres are left lying on the wire cloth and are carried forward as a wet fibrous layer. To distribute the fibres evenly, the wire is fitted with a shaker which agitates it sideways. The wet layer of fibre is then couched off the wire and passes, on a felt, through a pressing section, consisting of rolls which press the water out of the paper. The water which cannot be pressed out is then dried out in dryer cylinders in the drying section of the paper machine.

Even if no sizing were added at all, the fibres would still hang together after drying, forming a continuous sheet of paper with good strength properties. The fibres are said to have paper-forming properties, i.e. they are capable, after drying, of joining together with a certain force, enough to hold a sheet of paper together.

This all brings us back to what we have said previously concerning the structure of the cellulose fibre and the importance of the fibrils in the forming of the sheet. To make it at all possible to form an even and fairly regular sheet with acceptable strength properties, the pulp has to be dispersed in water, so as to reduce the fibre concentration to a few per cent, and forming then begins.

Without going into any further detail about the beating devices, we may note that the fibre suspension passes between cutters which slide over each other. Depending on the width of the cutters, their sharpness and their velocity in relation to each other, and also - perhaps the most important thing of all - the beating pressure, i.e. the pressure forcing the cutters against each other, the fibres will, in principle, be subjected to two different modes of treatment. One of them is cutting, which shortens the fibres, and the other is a certain crushing effect which, gradually, releases the ends of the fibrils, leaving them suspended like fringes round each individual fibre. This is particularly true of a textile fibre, with its coarse fibrils; in this case, the defibrillation process can be observed in an ordinary microscope. Parallel to defibrillation, the water-absorption capacity of the fibre increases and the de-watering of the fibrous pulp slows down. Extreme defibrillation produces a very strong, tough paper.

What really happens during drying is that the fibrils of the fibres come into such close contact with one another that they bond together. It is rather unclear whether these bonds are chemical or mechanical in character. Either way, this is the bonding force which holds the paper together and gives it its strength. It has also been found that, the greater the content of hemicellulose and saccharines, the greater will be the degree of maceration and the stronger the paper will become. To facilitate beating and to produce a stronger paper, organic substances are sometimes added which, in a solution of water, give a certain viscosity, e.g. quargum, CMC and suchlike.

If no sizing agents are added at all, one obtains a highly absorbent paper like blotting paper. To produce a paper which can be written on in ink, its absorption capacity has to be reduced so that the ink will not run. Sizing is also necessary to prevent printing ink from penetrating too deep into the paper or going right through it.

Resin sizing, which is the type of pulp sizing most commonly used, was invented by the German Illig in 1811. Before that, all paper had to be surface-sized with animal glue (gelatine or bone glue) in a special operation after manufacturing. The animal glue was hardened by adding alum (aluminium sulphate) to the sizing bath.

The introduction of resin sizing made it possible for sizing to be performed in the stock, simultaneously with beating or afterwards. This was a great advance and very important in the production of paper by machine.

The resin used is colophony resin (abietic acid). This has to be saponified first by cooking it in soda solution (or alkali), so as to produce the sodium salt of abietic acid. Papermills often buy ready-stabilised resin emulsions (bewoid size) for direct addition to the fibrous stock after beating. In order for the resin to be precipitated on the fibres and produce the sizing effect desired, the fibre suspension must have the right pH value and the resin must be precipitated as an insoluble metallic salt which is then deposited on the outside of the fibres. This is achieved by adding alum, so as to make the pH value 4.5-5.0 (acid sizing). When alum (aluminium sulphate) is dissolved in water, this releases sulphuric acid (by hydrolysis), whereupon the pH value falls and the aluminium ions precipitate the resin as the aluminium salt of the abietic acid, which is then deposited on the outside of the fibres. Later on in the manufacturing procedure, when the paper is dried on the drying cylinders, the resin is sintered and fills in the pores in the paper, giving it the requisite rate of sizing.

It is very common nowadays to neutral-size paper, i.e. to size it at a pH of almost 7. Sodium aluminate (pH:7.4) is then used instead of aluminium sulphate. Neutral sizing is an advantage because it gives the paper a very low degree of acidity, which slows down the degradation of its cellulose. Yellowing is also reduced. When sizing takes place in an acid environment, the resin itself can in certain cases contribute towards yellowing. Archival paper and other papers which have to be highly durable, as well as paper used for wrapping foodstuffs and certain machine parts, should be neutral sized. There are also special sizing procedures which are applied when the paper has to have very special properties.

Very often a wet-strength paper is needed, i.e. a paper which will not tear if it gets wet. In cases of this kind, resin sizing is combined with wet-strength treatment. This can be achieved by adding melamine resin or carbamide resin to the stock. Wet-strengthening can perhaps be performed in connection with the surface sizing of the paper, and paper can also be made water-repellent by adding wax or paraffin emulsions to the pulp.

Offset printing paper is frequently subject to the stipulations that it must not emit dust or contain fibres loosely attached to the surface, the reason being that dust and loose fibres adhere to the rubber blankets of offset presses, disrupting the printing. Resin sizing alone is not sufficient here, and surface sizing with starch in a size press is often resorted to. Starch solution (slurry), which can also be added to other synthetic sizes, e.g. "Basoplast", which increases the sizing effect, is rolled into the paper with a couple of press rolls, after which the paper is dried.

Surface sizing with gelatine or bone glue is done in a bath. The glue is dissolved in water and the web passed through the bath, whereupon the sizing solution is absorbed by the paper. The gelatine can be hardened in the same bath or in a separate one. Usually formaldehyde is used as a hardening agent, often together with other aldehydes. Alum (aluminium sulphate), previously used as a hardener, is very seldom used nowadays.

This surface sizing procedure is only used for textile fibre paper. It sizes very efficiently, at the same time greatly enhancing the strength properties of the paper. The result is a strong, durable surface with good printability, and this sizing procedure is particularly good for archival paper, bank note paper and certain types of drawing paper. Other surface sizing procedures, e.g. polyvinyl alcohol mixed with starch are also used. Acrylates and other special sizers are used as well.

When producing coloured paper, the usual procedure is for the colour to be added to the stock. Various types of colour are used - pigment colours, direct colours, acid or basic colours. Pigment colours, which have a high level of

Continuous sizing machine for animal size.

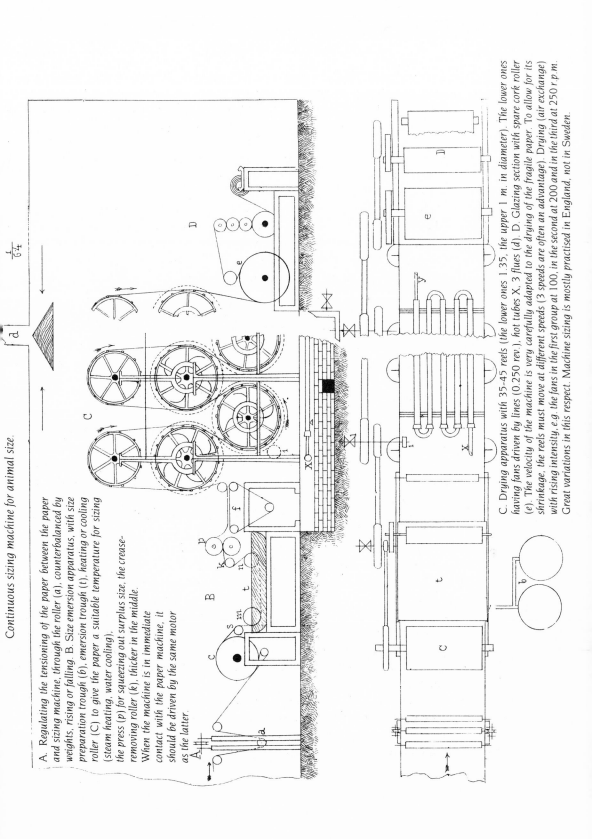

$\frac{L}{64}$

A. Regulating the tensioning of the paper between the paper and sizing machine, through the roller (a), counterbalanced by weights, rising or falling. B. Size emersion apparatus, with size preparation trough (b), emersion trough (t), heating or cooling roller (C) to give the paper a suitable temperature for sizing (steam heating, water cooling), the press (p) for squeezing out surplus size, the crease-removing roller (k). thicker in the middle. When the machine is in immediate contact with the paper machine, it should be driven by the same motor as the latter.

C. Drying apparatus with 35-45 reels (the lower ones 1.35, the upper 1 m. in diameter). The lower ones having fans driven by lines (0.250 rev.), hot tubes X, 3 flues (d). D Glazing section with spare cork roller (e). The velocity of the machine is very carefully adapted to the drying of the fragile paper. To allow for its shrinkage, the reels must move at different speeds (3 speeds are often an advantage). Drying (air exchange) with rising intensity, e.g. the fans in the first group at 100, in the second at 200 and in the third at 250 r.p.m. Great variations in this respect. Machine sizing is mostly practised in England, not in Sweden.

stability to light and very good resistance generally, are used for very high paper qualities. Alum is used as a mordant. Direct colours are also used for superior qualities. Acid and basic colours occur in simpler paper qualities.

Surface colouring is also used nowadays, because it is cheaper. With this method, the colour is applied to the paper by rollers. With this method, though, it can be hard to maintain a consistent colour.

The more the stock is beaten, the more translucent the paper will be. It is customary to say that paper is of low opacity. To improve this opacity, filler is added. Filler is used for quite a few kinds of paper, but the substances vary. Kaolin, a cheap filler, is often added to simple types of paper. Its coverage, however, is relatively small, and so quite a large quantity has to be administered. The drawback is that this impairs the strength properties of the paper.

For finer paper, titanium dioxide is used with the crystal modification anatase. Titanium dioxide is much more expensive than kaolin, but its advantage is its great coverage, which means that only a relatively small quantity has to be administered: 3-4 per cent is enough to make the paper sufficiently opaque. Nor does this smaller quantity have any appreciable effect on the strength of the paper. The most important thing in this context, however, is whiteness (lightness), which is essential to finer sorts of paper.

Various other fillers are used in the paper industry. Talcum or synthetic sodium aluminium silicate (Zeolex) is often used for very thin printing paper, e.g. Bible paper. Sodium aluminium silicate, it is true, has less coverage than titanium dioxide, but it is used together with the latter to prevent printing ink from running through thin paper. This is because sodium aluminium silicate has a very large specific area which absorbs all oil (printing varnish) in the printing ink before it has time to run through the paper. An organic filler based on formaldehyde has been developed in recent years. This is manufactured, for example, by Ciba-Geigy and is called Pergopak. It has proved efficient enough to compete successfully with titanium dioxide.

Glazing is a purely mechanical treatment of the paper and does not entail any chemical changes. Even so, glazing is so important that it ought to be mentioned, above all because it makes a big difference to the printability and general usefulness of the paper. Paper intended for letter press printing has to be highly glazed in order for the printing to be even, without any broken lines. Offset paper requires less glazing. The rubber blanket can apply ink even to a slightly uneven surface.

Copper-plate printing paper does not have to be glazed much. The main thing is for this paper to have a high bulk, so that it can be compressed under the heavy pressure it is subjected to in the copper-plate printer. In this way the paper will thoroughly penetrate the copper-plate engraving where the ink is deposited.

# Februarii Månad.

Fig. 1.

Fig. 2.

Fig. 3.

Fig: 1.

På Stenen a, hwilken står något lutande och är ganska slät, lägges hwart och et kort arck b. och medelst Stången c fram och åter förande, under hwilken Flint, Stenen sitter, Poleras hela arcket ganska jämt, och får man då ej ligga hårdare på någothera af flinthörnen, ty då blifver stora rifsor uti arcket. Fiädran d d som är ganska spänd gifver medelst sin Elasticitet Stången c, en stark tryckning. Det stycket hwaruti handtagen sittja ses Fig: 2 för sig sielf och är at taga lost, då flintan skall ömsas el: jämkas, är af Ek, Flintan tildess bredd och storlek så långt den kan ses utur trä. Stycket, synes Fig: 3. uti sina bägge Profiler, är af ordinair flintsten slipader, på den undre sidan.

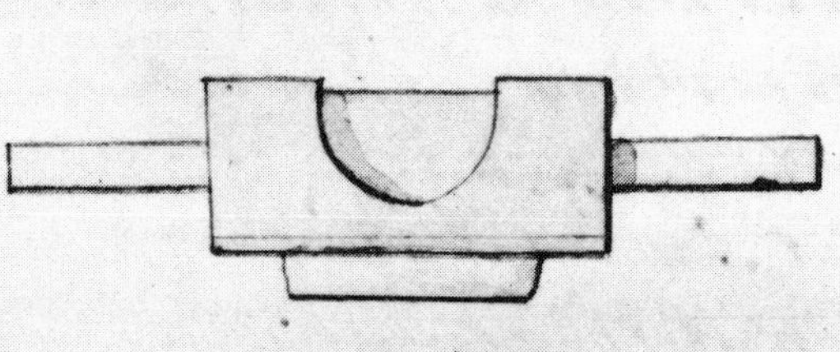

Innan arcket lägges under flintan at Poleras strykes en Sudd lost öfver et stycke Twål med hwilken sedan Arcket gnider, på det flintan skall därpå gå lättare, Sedan det således är Polerat blifver det färdigt at Klippas, då arcket först på 2ᵈᵉ sidor kanten afklippes, på en Sax a.

Stone glazing is presumably the oldest method for giving paper a hard surface. Sheets were already being glazed with agate or some other smooth polishing stone when papermaking was in its infancy. Only one side of the sheet was glazed, but because the sheet was placed on a wooden surface, the reverse side also acquired a certain finish. Glazing has the effect of blocking the pores in the paper and giving it a shiny surface which is more suitable for painting and writing with ink.

The earliest technique, quite simply, was to move a smooth stone backwards and forwards over the sheet. A time-consuming, tedious method which was eventually superseded by a technique in which the glazing stone was fitted in a holder but still moved manually to and fro. According to a diary entry in 1770 by Carl Bernhard Wadström, now in the Norrköping City Library (see previous page), the glazing procedure went as follows: "Each of the sheets (b) is deposited on the stone (a), which is slightly inclined and perfectly smooth, and by moving the bar (c), beneath which the stone is mounted, to and fro, the entire sheet is polished quite smooth, and one must not exert any extra pressure on any of the

*Sheet calender.*

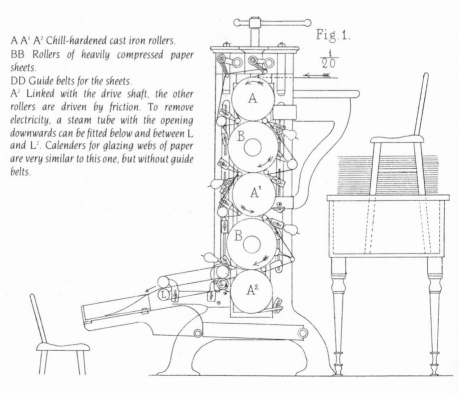

A A¹ A² *Chill-hardened cast iron rollers.*
BB *Rollers of heavily compressed paper sheets.*
DD *Guide belts for the sheets.*
A² *Linked with the drive shaft, the other rollers are driven by friction. To remove electricity, a steam tube with the opening downwards can be fitted below and between L and L². Calenders for glazing webs of paper are very similar to this one, but without guide belts.*

Fig. 1.
$\frac{1}{20}$

corners of the flint, because that will produce big scratches in the sheet. The spring (d), which is quite heavily tensioned, imparts heavy pressure to the bar (c) by reason of its elasticity. The section in which the handles are mounted is shown (fig. 2) separately. This is made of oak and has to be removed when the flint is to be changed or adjusted. The flint is wide enough and long enough to protrude beyond the wooden block - it is shown in both elevations in fig. 3 - and is of ordinary flint stone, ground underneath.

Before the sheet is put under the flint for polishing, a rag is gently wiped over a piece of soap, and the sheet is then rubbed with this to make the flint move more smoothly. After the sheet has thus been polished, it is ready for cutting."

This monumentally tedious job in turn was superseded by a simple machine which, on the whole, merely imitated the movements of manual glazing but was mechanically operated. The glazing stone was connected by a rod to a flywheel which moved it to and fro across the paper.

Another version was glazing the paper with hammers. This device was powered by water, usually by the same wheel as drove the stamping mill. The shaft was a wooden log, and the paper was placed under the actual hammer, which was of iron. A handful of sheets would be deposited on the plate, and by turning them during the glazing procedure one ensured that every square inch of every sheet received a number of blows. Simple hand-held hammers were also used. These resembled the planishing hammers used by bookbinders, and every stroke left its mark on the finished sheet.

In England, printing paper was already being glazed between two steel rolls in the 18th century, and eventually this method generated the next mechanical aid for giving paper a more resistant surface. This machine is called the calender, and one can distinguish between three main types of design. In type 1 the uncoated paper is put through alternate rolls of steel and compressed paper. The steel rolls are steam-heated. The steel and paper rolls are arranged in such a way that, during glazing, the web will butt equally hard against both, so as to make it equal-sided.

Type 2 presses the web between rolls while at the same time subjecting it to a certain amount of rubbing. This is achieved by putting the web through fairly tight rolls moving at different velocities. This type of calender is known as a friction calender.

Type 3 is used only for paper which is cut to size or hand-made, and the method involves pressing the sheet of paper, with a sheet of steel or zinc interposed, through close-fitting steel rolls. In everyday speech, this machine is known as a plater.

# WATERMARKS
# AND FILIGRANOLOGY

**Gösta Liljedahl on the use of watermarks as a means of identification**

*Previous page: Watermark, consisting of a crown and the monogram G(ustavus) A(dolphus) R(ex) S(ueciae). Uppsala, 1614. From Gustaf Clemensson, En bok om papper, 1944.*

Most people dealing with old paper in the form of written documents or early printed material know that, more often than not, it has watermarks, and several of them have probably also heard that these watermarks have proved a useful means of dating the paper itself and, consequently, its contents - the writing or print - which, of course, cannot possibly antedate the paper itself. Far too few people are aware, however, that watermark research, which has been energetically prosecuted ever since the turn of the century, has in recent decades undergone a notable renewal of methods, with the result that it is now far more adequate than it used to be as a "handmaid" of history together with such traditional auxiliary disciplines as diplomatics (the study of official documents) paleography, heraldry, numismatics. This has led to certain widely noticed results, as we shall now turn to see.

"Watermarks" in handmade paper are, quite simply, deliberate reductions in the thinness of the paper, in the form of contour lines forming letters of the alphabet or figures of different kinds. These are produced by sewing or soldering fine metal wires, previously fashioned in the shape required, to the wire cloth of the hand-mould (about which more below). When the mould has been dipped into the vat containing a gruel-like "stock" of pulp, lifted up again and skilfully shaken for a few seconds to make the fibres felt together more easily, while the water in the stock escapes through the wire cloth, the layer of fibres is retained as the finished sheet. Thus, at the points where the sewn-on wire figure is elevated above the base of the mould, the paper will be thinner, and the corresponding image will be visible in the sheet in the form of distinct, pale contour lines; the thicker wires, the paler or more distinct the "watermark" will be.

Watermarks also exist in machine-made paper, and these too are produced in basically the same way. Often the paper is thickened to produce darker shades, as for example in the "watermark portraits" in bank notes. The following account, deals exclusive-ly with handmade paper.

Why "watermark"? This is an ancient name in all languages. Clearly it was already provoking questions in 1708, when the New English Dictionary gave the following explanation: "The name was probably given because the watermark, being less opaque than the rest of the paper, had the appearance of having been produced by the action of water." In addition to watermark (vattenmärke, vandmaerke, Wasserzeichen, etc.), the Germanic languages used also to say "papermark". In the Romance languages, on the other hand, the dominant term is filigrane (Fr.); filigrana in Italian and Spanish), which is not a very apt term either, because it puts one more in mind of the "wire figure" than of the impression it leaves in the paper. This term has given rise to the word "filigranology", meaning the study of watermarks.

The watermark is a chapter in its own right in the history of paper-making, and, obviously, space here will not permit more than a very concise account of the subject.

Paper in our sense, i.e. made from matted vegetable fibres, was invented in China, probably in about 100 B.C. Curiously enough it took something like 1,000 years for the jealously-guarded secret of paper-making to reach what is now the western world; travelling through the Arab world, where it became known in the mid-8th century, it arrived in Spain, at that time governed by the Moors, in the 11th century. Not until the mid-13th century do we find any signs of paper being made in non-Spanish Europe, viz in Italy. The first recorded papermill in Italy was in full swing in Fabriano, in the province of Ancona on the eastern slopes of the Apennines, in 1276. Possibly this little town produced the first watermarks, as a simple means of distinguishing between products from several papermakers who were active there - no fewer than twenty in 1330.

The question why this practical invention was not made earlier by the Chinese or the Arabs is a fairly easy one to answer. It seems that the attachment of an emblem to the mould had not been technically feasible until now. Starting probably in Spain, but definitely in Italy, several important improvements had been made to paper-making techniques, e.g. by introducing a solid mould - a wooden frame with a bed of iron wires - whereas Oriental paper-makers had always used - as they still do - a soft mould made of narrow strips of bamboo joined together with strands of hair or silk. Wire figures could not be permanently attached to this pliable mould, in contrast to the Italian construction.

It has been suggested that the first watermark came about by accident. A metal wire on the bed of the mould may have come loose and made an extra, unintentional impression in the stock -not exactly an uncommon event - and some canny individual may have realised how a "mark" of this kind could be developed and made use of. However, this may be, after the first recorded instance in 1282 (a cross), a whole flora of watermarks quickly developed in Fabriano and elsewhere in Italy, starting with rudimentary symbolic characters - crosses, circles, stars and so on - and eventually graduating to primitively fashioned animals, plants, human figures, artefacts used in paper-making, and so on. The manufacturer's name was also used (beginning in 1305). Most medieval watermarks no doubt related to a Christian repertoire of symbols which everybody could easily understand then but which today are more or less a mystery, the meaning of which can only be guessed at.

*Hand-made watermark from Tumba Papermill, with both reductions and augmentations of the paper thickness used to produce a shaded "portrait" effect.*

From Italy, which for a long time remained the leading "paper country", with a flourishing industry and copious exports, the new "art" spread across Western and Central Europe to England, Eastern Europe and Scandinavia. The use of paper, of course, was widespread in all the various countries even before they began producing their own paper. In Sweden, for example, paper had been competing with parchment ever since the late 14th century, but Sweden's first papermill, a royal mill at Norrström, was not founded until 1565. (It was closed down after a decade or so and only one watermark is known from it - predictably, a "three crowns" emblem. The next papermill - this time in Uppsala - was not started until 1611.)

Practically all "superior" paper was watermarked. Production in Italy was subject to fairly strict regulations, and we know for a fact that the practice of giving the paper a watermark soon came to be looked on as a means of preserving its quality. Plainer, "rough", paper - with or, most often, without a watermark - was probably also manufactured early on, but it is impossible to say in what quantities, because very little has survived.

The other countries followed Italy's example, and the watermark came to be widely regarded as a quality mark, even though the quality was not always

215

exactly first rate. Imitations of well-known goodwill marks naturally abounded, they were forbidden in constantly reiterated ordinances attended by penalties and legal proceedings, but clearly to little avail. This causes quite difficult problems of provenance, but the big problem in watermark research is that certain marks clearly lost their character of factory marks altogether and became quite "free", exclusively designations of quality - in some cases they were also attached to different formats of paper - and as such they could be used by any papermill whatsoever. This happened, for example, between 1325 and 1600 with the familiar, originally Italian "Bull's Head", in which the bull, an ancient symbol of strength and perseverance, doubtless came to denote a strong, durable paper, and as a quality mark of this kind it enjoyed uniquely longlasting popularity. A German scholar, Gerhard Piccard of Stuttgart, has identified, copied and dated something like 24,000 variants (I shall be returning presently to this vital but rather puzzling concept) of this watermark alone for the period 1325-1650, and he has organised his material into about 4,000 "types"! The Bull's Head definitely tops the list, thanks to its long life, but Piccard himself says that intermittently during the period it was probably surpassed in frequency by several other watermarks, e.g. the "Gothic" letter P (the interpretation of which is a matter of dispute), the numbers of which are yet to be calculated.

As paper-making expanded, new watermarks were continuously added and the thematic repertoire was growing all the time. For example, watermarks with heraldic themes, armorial shields showing the bearings of the aristocratic owner, the royal patron or the manufacturing centre, were very popular indeed. But these too were liable to be forged and - more or less innocently - imitated, and indeed sometimes they became entirely "free" watermarks, as for example in the case of the city arms of Amsterdam (watermark c. 1660-1780), which, beginning of course as a Dutch watermark, soon became current in every country, Sweden included.

It has been estimated that, merely in the period down to 1600, more than 150,000 different watermarks were probably produced (and probably more than a million by 1800!), by far the greater part of them consisting of a relatively small number of motifs which, in the manner already described, were used as "common" marks. As we have already seen, the use of a certain mark to denote prime quality was by no means a full guarantee in this respect, and both responsible manufacturers and authorities with an eye to the reputation of their country or locality struggled to achieve better control of watermark emblems.

The growing tendency to add one or two additional marks to the "free" watermarks, as a more exact specification of origin, is looked on as a precaution of this kind. To the simple contours of the "Bull's Head", for example, there would often be added a rising "staff" between the horns, supporting a symbolic object - a cross, crown, flower, wriggling serpent, letter of the alphabet etc. -

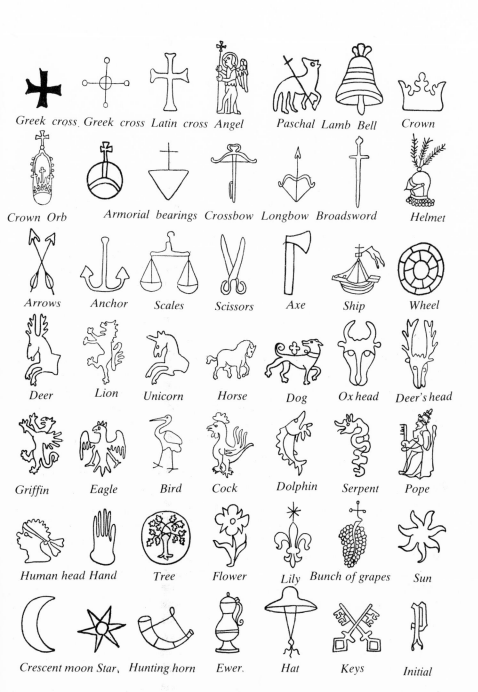

| | | | | | | |
|---|---|---|---|---|---|---|
| Greek cross | Greek cross | Latin cross | Angel | Paschal Lamb | Bell | Crown |
| Crown Orb | | Armorial bearings | Crossbow | Longbow | Broadsword | Helmet |
| Arrows | Anchor | Scales | Scissors | Axe | Ship | Wheel |
| Deer | Lion | Unicorn | Horse | Dog | Ox head | Deer's head |
| Griffin | Eagle | Bird | Cock | Dolphin | Serpent | Pope |
| Human head Hand | | Tree | Flower | Lily Bunch of grapes | | Sun |
| Crescent moon Star | | Hunting horn | Ewer | Hat | Keys | Initial |

*Initial types of watermark used in the Middle Ages. (Greatly reduced). From Briquet, Les filigranes.*

217

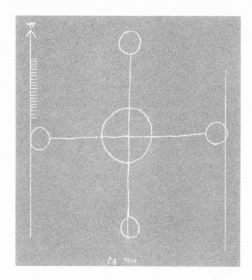

*The oldest known
watermark, in a
document from Bologna,
1282. (From Les Filigranes.
110 x 108. Reproductions are
on a reduced scale, unless
otherwise indicated. The figures
refer to the maximum height
and breadth of the original in mm.)*

which in turn could assume a wealth of different guises. Since, however, the proliferation of such additional watermarks developed into a jungle and lost its original character - many of these marks too apparently became quite "free"! - initials, usually those of the owner or lessee operating the papermill at the time, came to be added instead. Presumably these latter additional marks worked quite well as factory emblems, and from the viewpoint of the watermark specialist, of course, they represent a great improvement, even though today we can only interpret a fraction of them.

In time, of course, there came to be a tremendous diversity of initials, and interpretation is not made easier by a growing tendency to use three or more instead of just two. The initials used were not only initials in the strict sense: certain letters would be excerpted from names of persons and places, titles and other words, to form long, frequently enigmatic combinations. Here are some examples, the interpretations of which are fairly reliable: IHVBS = Johann Ulrich Beckstein, IHLAW = Johann Lorch auf der Annweiler Wachtelsmuhle, FFGUHVP = Friedrich Ferdinand Graf und Herr von Pappenheim, WEHZSICBEVW = Wilhelm Ernst Herzog zu Sachsen, Julich, Cleve, Berg, Engern und Westfalen! Sometimes the initials would be reversed (FB = Bartlin Ferber, HD = David Hess), sometimes they would be haphazardly re-arranged, so that we find Jörg Muller of Annweiler, for example, variously styling himself IM, MI, IMA and MIA (added to which these letters, like a number of others, can of course be read from both sides - that is, how you read them depends on which side of the paper you are looking at).

By the end of the 17th century, initials began to be applied to the other half of the sheet, which until then had usually been "vacant", which explains why scholars often failed to find any watermark in the documents they are studying; not all documents are whole sheets. In this case one distinguishes between the main watermark in one half of the sheet (usually the left half, but very often the right) and the countermark in the other. The "general" watermarks which now occur - quite different from the previous ones and mostly Dutch, e.g. the "arms of Amsterdam" and the immensely popular "Pro Patria" - almost invariably comply with this rule, while for example the jester's head (or rather, "Foolscap", now surviving as the designation of a format) generally undergoes a process of transition between 1640 and 1690 and does not begin to acquire a countermark until the 1670s.

To begin with, no distinction was made between writing and printing paper; incunabula then, were printed on ordinary writing paper with watermarks to match, and the same goes for many later books, especially those in folio format and of a more pretentious kind. Printing paper already began to be left unsized in the 16th century - to stop the ink running - and eventually less high-quality grades of paper, frequently without watermarks, began to be used for everyday printing. When watermarks do occur, however, they are often quite difficult to discover and identify in smaller formats than folio, due to the folding and cutting of the sheet. In folio they usually appear vertically in the centre of one half of the sheet, in quarto they occur horizontally in the middle of the fold between two sheets (1 and 4 or 2 and 3), in octavo central parts of the watermark have been cut away, and more or less mutilated "outer" quarters of the watermark occur vertically in the top inner corners of four sheets (1, 4, 5, 8 or 2, 3, 6, 7). In smaller formats still, fragments of watermarks can be discovered and can be of interest to the bibliographical researcher, even if the watermark itself is unidentifiable.

To understand the conditions governing true watermark research, we must return to the mould and its technical design. It remained fairly unaltered throughout the era of handmade paper in the western world, consisting of a rectangular wooden frame with a number of longitudinal slats 2-3 cm apart. The frame matched the format of paper to be produced, so that every format needed its own mould. On the slats there rested a "cloth" of metal wires, stretched at fairly close intervals from one end of the mould to the other. These are called "vergé wires" from the French vergé = striped. Two thinner metal wires - known as "warp" or "chain" wires - were stretched over each of the slats at right angles to the vergé wires, through which they were woven, so as to hold them together and keep them in position. The warp wires are so close together that they look like just one wire; sometimes, indeed, there was only one - slightly coarser - wire. The vergé and warp wires together form the "wire cloth" or screen, which was attached to the slats by very fine metal wire. The "wire figure" could then be sewn or soldered onto the screen. A detachable, close-fitting wooden frame, the "deckle", was positioned on the mould to keep the stock from running off the mould altogether.

A sheet of handmade paper, quite simply, is a pulp cast of the screen, and the various wires in it each leave more or less distinct impressions in the sheet - the layer of fibre is made thinner above all the various elevations. The vergé wires produce vergé lines in the paper, the warp wires produce warp lines, the wire figure produces a watermark and, finally, the attachment wires and sewing threads leave their attachment marks and sewing dots (the two latter impressions being often indistinguishable). All these units have to be carefully compared for identification purposes, and of course the same goes for the important, easily observable detail of the positioning of the watermark in relation to the warp lines. Complete scrutiny of this kind also makes it possible to perform the very important task of identifying "mould pairs", i.e. the two mould halves, belonging together, which, for practical reasons, were nearly always used in paper-making.

When the "vatman" had lifted up one mould from the vat of stock, shaken it and formed a sheet, he removed the deckle and pushed the mould across a plank to the "couchman", who, with a rolling movement, deposited the sheet on a woollen cloth (a "felt") - "couched" it (from the French coucher). Shortly before that he had pushed the other mould, recently relieved of its sheet, across to the vatman, who now fitted the deckle to it and moulded a new sheet, and so on, with the two moulds passing regularly to and fro between the two men in a steady and efficient rhythm. The couched sheets with their felts in between were collected into a suitably sized pile ("post") which was then pressed twice, once with and once without felts, to extract most of the water from it, after which it was sent for further processing, e.g. drying, sizing, smoothing and sorting.

These two moulds with their single deckle were made simultaneously, usually by a specialised craftsman, the mould-maker. The screen with its vergé, warp and attachment wires could be made almost identical for the pair, but whether, as in earlier times, he bent the figure wires with pliers or, as later came to be the practice, used a template (often a block of wood incised with the drawn contour lines. The figure wire would be pressed into the grooves whereupon the finished figure would be lifted away and transferred to the screen), he could not with the best will and all the dexterity in the world produce an exactly identical figure; at least when he came to attach it to the screen, if not earlier, some difference compared with the system mould was bound to result. Perhaps too the figure would be slightly differently positioned on the screen, and the sewing points for the figure wire did not always match completely.

If, then, two sheets whose watermarks differ only in minor details present exactly similar screen impressions, we are dealing with a pair of moulds and the watermarks are "twins". In a manuscript or book one can often follow them page

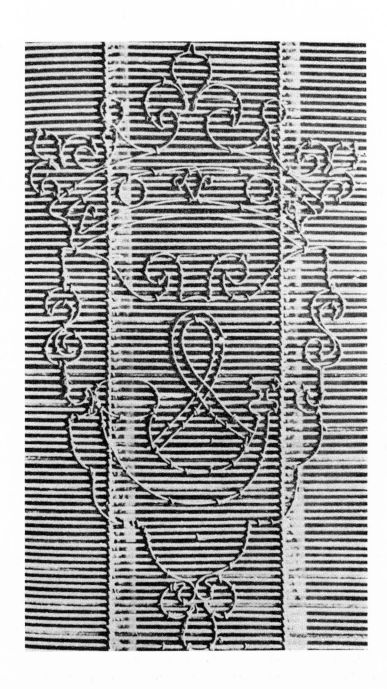

*Detail of a hand-mould, showing vergé and warp wires and the wire figure. From H.
Voorn, De papiermolens in de provincie Noord-Holland, Haarlem 1960.*

*Uddby in the parish of Tyresö, founded in 1620 by Gabriel Gustavsson Oxenstierna, was one of Sweden's first papermills. The family crest, variously depicted, made a decorative watermark. This "pair of twins" dates from c. 1640. Note the many small but perfectly distinct differences of detail! (a) 88 x 75, (b) 87 x 76, Swedish National Archives.*

by page - a-b-a-b-a-b and so on, as they came in the "post". Of course, the regular alternation is broken every now and again, partly because one or two defective sheets have been rejected, and partly because the order has been disrupted in the course of drying, sizing and so on; but there are always two alternate marks.

In principle, then, one should immediately find the pair if one has two sheets of the same kind of paper (a-b), but as we have now seen, "disruptions" are possible, and so several sheets have to be gathered together to establish this point. The larger the number of sheets of the same sort, the better chance we have, then, of establishing the pair of moulds. In account books, for example, where the sheets are bound in gatherings, the paper is often uniform in long sequences, providing a full illustration of the twin relationship. But shorter sequences are also good enough, right down to the smallest possible, a-b, therefore for "complete proof", of course, one tries to find more ample evidence. Sometimes, of course, an individual document will only feature one of the twins and the other will never be found, and often the two occur in widely different places, in which case the best one can do is to surmise that they belong together.

Even though this almost invariable working technique, using two moulds, as similar to one another as possible, conjointly, was well known to specialists, earlier watermark researchers, unfortunately, did not draw the right conclusions. For the great significance of the practice to research is that the frequently enormous number of variants of a watermark can be divided into pairs, each in use for the same length of time; in other words, the confusing abundance of data can be reduced by exactly half! (So much for theory; in practice, as usual, minor complications occur, but these can be disregarded for present purposes.) And once the pair has been identified, then for dating purposes it is sufficient to find one twin or the other. In other words, even individual documents, consisting of a whole sheet or half sheet or perhaps of just a slip of paper with a reliably identifiable fragment of a twin, will then be good enough for identification purposes.

This brings us on to the core issue: the durability of the mould and the wire figure. Experimentally it has been concluded that a normal handmade paper mould can be used to produce about 500 reams of paper (one ream = 500 sheets); it will then be worn out and will have to be renewed. Since the average annual output per vat and pair of moulds in continuous operation was about 1,500 reams, i.e. 750 reams per mould, this would give a pair of moulds a useful life of about eight months. In practice, though, a pair of moulds was seldom used continuously. For one thing, every papermill, large or small, would use several pairs of moulds -different pairs for different formats - in the course of a single year. Most mills were small and had only one vat, which meant that they could only use one pair of moulds at a time. And then there were liable to be disruptions of many different kinds: fires, war, shortage of materials (a shortage of linen rags was a perpetual bugbear in all countries), but above all it was lack of water power that caused many stoppages, during both dry summers and cold winters. Taking all these various things into account, a true service life of about two years would seem a more realistic assumption. This applies to a mould used for making ordinary writing paper; moulds for less common grades or formats, of course, have the same true service life (8 months) but were actually in use for a good deal longer - between 5 and 15 years is the usual assumption.

The wire figure, of course, was the most fragile part of the mould, the first part which needed replacing. The stitches came undone, the contour wires were displaced, the whole figure could slip in one direction or the other. It was secured once again and rough and ready attempts were made to reconstruct it, but as a rule it was left in position on the screen until the latter was in such a poor state that the entire mould had to be replaced (in exceptional cases, one twin had to be completely replaced before the other one was used up, or again, both twins might be completely replaced on the old screen; cases like these, which are

*Different states of the same watermark. CF = Cathrinefors Papermill (c. 1825). The F. stays
the same, but the C work loose, slides on the screen and is slightly distorted. Actual size.
Swedish National Archives.*

probably quite rare, are among the inevitable complications we have just hinted
at). The ongoing deterioration of the wire figure is reflected by the watermark,
which nearly always displays distinct "stages" in its life, thus giving rise to a
number of "variants" of the same watermark. These "states" ("deformed
variants", "phase variants") can be quite different from each other, but they are,
quite rightly treated as one and the same watermark and therefore also known as
"false variants", to distinguish them from the "genuine" variants, meaning the
other twin (including its states!) and also similar marks from other pairs of
moulds. Genuine variants can be designated a-b, c-d, e-f and so on, states,
respectively, a1, a2, a3 etc., b1, b2, b3 etc., and so on.

The states for a certain watermark are usually quite easy to distinguish from
the states of the twin, and also from other genuine variants, because there nearly
always remain a sufficient number of absolutely distinctive details from "stage
1". For example, ten out of twenty distinct characteristics may have become
unrecognisable, but the remainder are more than sufficient for reliable
identification.

The states could also be called "identical variants", but that term has
previously been used in a vague sense (by Briquet, see below), mostly to denote
"very close similarity", and it ought preferably to be dispensed with.

Theoretically, then, any watermark whatsoever in ordinary writing or printing
paper should be datable to within about two years, its normal useful life. Is dating
of this accuracy possible in practice as well, and if so, how is it done? The year
of origin, at all events, can be pinned down, but the year when the wire figure
ceased to be used cannot, as a rule, be specified with the same degree of
confidence. Since direct information from early papermills on the service periods
of their watermarks is very rare indeed - records of litigation and suchlike
provide rare opportunities of this kind - comparison with reliably dated, identical
material is practically the only possible recourse. In documents from a suitable
time and place (an archive researcher or bibliographer can usually judge the

approximate period by other criteria), one looks for identical marks, as many as possible, preferably from different archive materials. Suppose we are lucky enough to find 20 identical copies. Perhaps three of them occur in various documents from 1673 (starting in August), 11 from 1674, three from 1675, one from 1676, one from 1678 and, finally, one from 1681. We conclude, of course, that the paper should have been made during the latter half of 1672 or possibly at the beginning of 1673 - after all, it took some time for the paper to come on the market, and in the case of imported paper, nearly a whole year could elapse between production and use. As can be seen, we obtain, with quite considerable certainty, an initial date, a "terminus post quem" (or more drastically: "ante quem non"), prior to which the watermarking question was not used, which means that no earlier document can exist having this watermark.

The final limiting point in time for the use of the watermark, its "terminus ad quem" ("post quem non"), can, as we have already remarked, not be determined with the same degree of certainty, even though the distribution of the currencies in our example suggests that, here at least, the two-rule holds good. But by no means all occurrences present such a "neat" and unambiguous graph line, and, as ever, it is the exception that proves the rule.

The terminus ad quem of a document incorporating the watermark in question is, of course, a matter of much greater uncertainty; the writings/print may, of course, be of a much later date. Statistical analysis has shown that the great majority of newly manufactured paper was consumed within a few years, but probability assumptions of this kind are little use in the individual case. But the possibility of fixing one limit is something to be grateful for and it is more than other "auxiliary sciences" can usually achieve!

As will be readily understood, this comparison with dated material is an exceedingly time-consuming, painstaking business. Merely finding a sufficient number of instances involves no small measure of difficulty, and the actual identification of one watermark with another is a very intricate undertaking, even for the specialist. As watermark research advances, however, our accumulated knowledge of this field grows accordingly, and new findings are being published all the time. All that can be offered here are a few indications of the aids available in the form of literature and other things.

Filigranology as a "science" can be said to have existed since the end of the 19th century, and as such it is above all associated with the name of Charles-Moise Briquet. The early publications on watermarks are now mainly of interest as documents in the history of scholarship. Quite naturally, a good deal of their content is outmoded and partly inaccurate, and those conclusions which have stood the test of more modern research have been incorporated in more recent writings.

Briquet had published a good deal on the subject of watermarks before 1907, but that year he published, in Geneva, his massive "Les Filigranes. Dictionnaire historique des marques du papier dès leur apparition vers 1282 jusqu'en 1600 avec … 16112 fac-similés de filigranes", which instantly made him a household name in the world of scholarship. This magnum opus was reprinted in Leipzig in 1923 and a slightly revised edition was brought out in Amsterdam in 1968. Briquet's 16,000 watermark reproductions were merely a selection from about 44,000 drawings, but data are also included for the 28,000 variants which, for reasons of economy, could not be reproduced. Not the least valuable aspect of the book was - and remains - Briquet's introductory comments on each watermark and principal type.

The impression gained in 1907 was that all watermarks from the period in question were represented here, so that watermark research was "concluded"; all that possibly remained was to add or polish a few details. Today we know that Briquet's 44,000 registered watermarks are only a more or less random sample of types and variants, and we can understand why a researcher consulting Les Filigranes seldom is able to find the exact counterpart of "his" watermark. Briquet, for example, depicts 1,363 "Bulls' Heads" and mentions the existence of a few thousand other variants. Piccard has confirmed more than five times that number, and there are probably quite a few more!

Briquet was a researcher of quite different stature from his predecessors and contemporaries in the same field, and his merits are not diminished by our ability today, more than half a century later, to perceive defects of method in his magnum opus. First and foremost, he makes the mistake of not arranging his watermarks in jointly dated pairs of twins. Briquet was perfectly well aware of the technique of using two moulds, but, oddly enough, he never realised the filigranological importance of this fact (as he himself later deplored in a letter to the German watermark expert, Karl Theodor Weiss, who was the first to observe this fact and to draw the important conclusions from it). As a result, in order to instil some semblance of order in the plethora of different variants, he was obliged to group them under the vague, sometimes confusing designations "varietés identiques", "v. similaires" and "v. divergentes", a basis of classification which he was not really satisfied with himself and which his readers could never quite make sense of.

The other drawback with Les Filigranes - it could not be termed a mistake at the time - is that Briquet, quite naturally, made tracings of his watermarks, which today is not considered an absolutely dependable method. It is suspected that his reproductions, although doubtless very painstakingly done, do not always tally with reality; their lines are just a little too confident and unhesitating to be absolutely trustworthy! The view today, and quite rightly so, is that it is immensely difficult, indeed practically impossible, for a simple drawing to reproduce with complete accuracy every significant detail of what is sometimes a very indistinct watermark; the draftsman "misses" some points and instead he draws non-existent lines which he believes himself to see. In Briquet's day, hardly any other method of depiction was conceivable. True, he does mention photography as a desirable method, but he rules it out because of the prohibitive expense. A satisfactory solution to this problem is absolutely vital, and I shall soon be returning to the view taken of the matter by modern research.

Briquet's immense authority and the almost paralysing sense of there being nothing left to say about watermarks were if anything detrimental to nascent watermark research. It stagnated and did not get under way again until the mid-1930s. Today it is being conducted with great vigour and with vastly improved methods.

First and foremost, it has been realised that the persistent hope of the filigranologists to depict and publish all watermarks is sheer wishful thinking. It has been estimated that watermarks down to 1600 alone would require more than 50 folio volumes the same size as Les Filigranes, and so this project has been dismissed as completely unrealistic. Publication of watermarks, therefore, has partly - and, understandably, as previously - been naturally linked with general research into the history of paper, including the history of manufacturing centres and personalities, or again, in more specialised filigranological works, an attempt has been made to delimit and at least to treat with some thoroughness watermarks from different regions and localities. For instance, the Paper Publications Society in Hilversum (now Amsterdam), founded by a British immigrant, E.J. Labarre, has done very creditable work on these lines, so far publishing twelve majestic folio volumes in its "Monumenta" series, together with a number of incidental publications, including the new edition of Les Filigranes mentioned above. These works are excellently edited, they have first-rate indexes and so on, but their method of depiction is conventional drawing, added to which one wonders whether even this more sober objective is feasible or worth maintaining in the long run. The material thus published is growing more and more daunting in volume, great expense is involved (though it would be greater still if authors and associates were not paid so little), and the available resources could perhaps be put to better use.

A more viable and sensible way seems to be the establishment in different countries of central institutions for national research into the history of paper - which of course must also devote considerable attention to the important subject of imported paper - with regular "customer services" for researchers in need of assistance. Institutions of this kind already exist here and there, e.g. in West Germany and the Netherlands: "Forschungsstelle Papiergeschichte" in Mainz and "Stichting voor het onderzoek van de geschiedenis der nederlandse papierindustrie" in Haarlem, both founded (in 1939 and 1953 respectively) and maintained by the paper industry, and also in several Eastern European countries, where they have been set up as an integral part of state cultural activities, forming special departments of academies of the historical sciences.

The Nordic countries too have at least the embryos of such research centres, viz one in Helsinki, affiliated to the central laboratory of the paper industry, one in Copenhagen, affiliated to Det kongelige Bibliotek, another in Oslo, affiliated to the University Library, and a fourth in Stockholm, affiliated to the National Archives. Interest in the activities and development of these institutions shows encouraging signs of prospering. The twenty-year-old Association of Nordic Paper Historians is campaigning vigorously among the appropriate authorities and agencies to promote understanding for its wishes and viewpoints.

The furthest headway in the important matter of customer services would appear to have been made by Hauptstaatsarchiv Stuttgart, where the paper historian Gerhard Piccard, mentioned previously, is active. In 1961, acting in close collaboration with the archives, he began publishing a series of "Findbucher", each dealing with one particular watermark motif, the intention being to treat it exhaustively. Only a selection of "types", however, are reproduced in a Findbuch; the basic material, in Wasserzeichenkartei Piccard in the archives, consists of about 100,000 pencil drawings by Piccard, which are successively being transferred in indian ink to sheets of cardboard. The intention is for a researcher interested in a particular watermark to try and find the right type, obtain information about its occurrence etc. and then, by sending Kartei Piccard an exact drawing, to get his watermark individually dated. The two works hitherto published in the series planned deal with "Kronenwasserzeichen" between 1385 and 1672 (5,000 watermarks drawn from 3,800 documents; the book shows 547 types) and "Ochsenkopfwasserzeichen" from between 1325 and 1650 (in 3 volumes; of something like 24,000 watermarks drawn, 3,993 are reproduced as types).

The idea itself is excellent, but its implementation has come in for criticism. Piccard is an outstandingly knowledgeable paper historian and doubtless a skilful, seasoned draftsman, but his adherence to a method of depiction which must be considered far too inexact arouses misgivings, and all the more so his total disregard both of the "twin" relationship and the appearance of the actual screen; in his view, the individual watermark is the only thing that counts for purposes of comparison and dating. It would be going too far here to describe the reasons he advances for this startling departure from generally accepted "doctrine" (they are set out, for example, in his introduction to "Ochsenkopfwasserzeichen") and the dispute to which this has given rise. Clearly, though, there are fundamental points on which Piccard stands alone against unanimous expertise, which greatly detracts from his authority.

Every auxiliary science requires methodological guidance, preferably codified in manual form, and filigranology today has two such works at its disposal. The

*Mould with light-and-shade watermark. From Hunter.*

more exhaustive of the two, published in 1962, is, very appropriately, entitled "Handbuch der Wasserzeichenkunde". Its original author was Karl Theodor Weiss, mentioned earlier, who died in 1945, and since then it has been essentially revised and augmented by his son Wisso Weiss, the DDR's foremost expert on the history of paper. The subject is treated with what one is tempted to call typical German thoroughness, giving the reader the impression of scrutiny from every conceivable angle - historical, methodological and purely scientific. But one looks in vain, for example, for a bibliography, though its absence is explained by this manual being but the first part of an intended series, the remaining three volumes of which, in addition to the bibliography, would also include complete lists of all known papermakers and papermills. If this work of reference can be completed according to plan, it will be a very great help indeed to paper historians.

The smaller work, entitled "Datieren mit Hilfe von Wasser-zeichen", was published in 1964 by the late Theo Gerardy, an eminent paper historian in Hanover. This is not the place to analyse "Methode Gerardy", as he calls it, a method which he eagerly advocates. Suffice it to say that, as an engineer and "Oberregierungsvermessungsrat", he endeavours to apply mathematical accuracy - based on careful measurement of details of the whole sheet of paper, watermark included - theory of errors and probability calculations so as to impart scientific stringency to the study of filigranology. Above all he maintains that,

229

using a codified description of essential units, with the addition of all provenance data etc., one can achieve, in tabular form, a systematic catalogue of watermarks which, in terms of convenience and lucidity, will greatly surpass a captioned pictorial catalogue of the Briquet type, leaving aside the prohibitive cost of the latter. The code can very well be combined with reduced-scale depictions of watermarks, as well as schematic sketches of essential details of the variants, so as to elucidate the language of the mathematical formulae when necessary.

In terms of principle, Methode Gerardy is unexceptionable. It goes without saying that a general catalogue of watermarks - in itself a natural objective in order to make filigranology a fully serviceable auxiliary science - must include all essential data. And indeed, the debate on the method recommended is concerned not so much with the principle as with the extent to which it needs to be applied, which data are really "essential". Wisso Weiss feels that Gerardy's objective - the mathematically compiled catalogue "mit allen Rafinessen" - demands such an input of time and labour that, for the moment at least, it should be put aside in favour of more urgent tasks which can be completed within a reasonable space of time. The careful systematic division of watermarks, aimed at reliably positioning any watermark whatsoever in the system, and the rapid tracing of the watermark in the finished collection -these, Weiss maintains, are the essential requirements, and more rudimentary registers and indexes are quite good enough.

Gerhard Piccard contemptuously dismisses Methode Gerardy in one word: "Perfektionismus!"

To an onlooker, observing the ongoing debate between "the prophets" with all modesty but also unburdened by prestige, both Methode Gerardy and the various classifications systems recommended - each of the protagonists has a carefully worked out system of his own - all seem to have their advantages and it is hard to tell which of them is "best". Since they all appear to work well, the truth, here as on so many other occasions, is probably that each of them is the right and natural system for whoever has arrived at it by dint of prolonged experimentation; other researchers can opt for either system or, via alterations and modifications, create systems of their own. A uniform international system would of course be desirable. It would confer tangible benefits, but at present it does not seem to lie within the bounds of possibility. Co-operation is thus rendered that little bit more unwieldy, but it cannot be helped.

Otherwise the necessary international co-operation runs smoothly, apart from the rather sluggish nature of communications with the countries of the Eastern bloc. "Internationale Arbeitsgemeinschaft der Papier-Historiker" (IPH), founded in Mainz in 1960, includes, it is true, a number of members from the Eastern European countries, but researchers there tend more often to belong to

*a-b) Wideface - Squareface, c-d) Leveleye - Slanteye. Bull's Head twinpair from Missale speciale c. 1473. Natural size. From A. Stevenson, The problem of the Missale Speciale. London 1967.*

and act within their national organisations only, and they frequently have difficulty in visiting the regular IPH congresses. IPH's well-filled membership journal, published in Ba Velp, Netherlands, by Henk Voorn four times annually, and the journal "Papiergeschichte" (PG), published six times annually by Forschungsstelle Papiergeschichte in Mainz, are good inter-mediaries.

Of vital concern to anybody dealing with watermarks is the question of how best to depict them. We have already remarked on the unreliability of manual drawing, but what substitute is available? Firstly, it should be stressed that there is still a place for the simple, inexpensive drawing in cases not demanding absolute accuracy, e.g. initial comparisons or illustrations. In more demanding contexts, one can choose between four other expedients (specified here without any closer description, merely with a few brief remarks).

1. Conventional photography with a miniature camera provides a good picture of both watermark and screen, is fast and simple but is not all that cheap if the negatives are to be enlarged again to actual size. 2. Photostat copying, using diazo-film, is also fairly simple and much cheaper, shows the original actual size and with good resolution, but has certain minor drawbacks.

Xerography requires expensive apparatus and have been practically tested to such a little extent that a layman would not dare to pronounce on their usefulness. 4. Beta spectrometry, i.e. radiation with a radioactive isotope (Ca 45, C 14 or Pm 147), originally by the Erastov method, which was first devised in Leningrad. Beta spectrometry is a more expensive and slower process, but it is clearly superior to the other methods in the important sense of quite eliminating the perpetually obtrusive text/print on the paper (printing or other ink being,

231

normally, such a thin surface coating in relation to the thickness of the paper that the radiation passes through it without the photographic paper recording any "extra" mass density).

Librarians and archivists, quite naturally, are undecided concerning this new procedure, but it has now been tested for such a length of time, with positive results, that "destructive" hazards to both human beings and materials seem to be excluded. In an article published by PG in 1967 (No. 32), Ove K. Nordstrand, Conservator at Det kongelige Bibliotek, Copenhagen, has specified six laboratories in Europe - one of them at the British Museum -where experimentation has been in progress for various lengths of time, and the method is now being employed on a routine basis in more places than one.

The filigranologist who has most assiduously employed beta spectrometry in his investigations is the American Allan H. Stevenson, one of the foremost experts in the "analytical bibliography" so eagerly cultivated in the USA and Britain. Already in the late 1940s he made important contributions to the discussion, in progress ever since the beginning of this century, concerning the dating of Shakespeare's quarto editions, and when the Erastov method became known in about 1960, he was one of the first people to put it to practical use. Two of his research findings which caused quite a stir in specialist circles - his datings of "Missale Speciale" and of the "Vinland Map" - will be briefly related here.

Missale Speciale Constantiense was at the turn of the century the only known copy - disregarding a couple of fragments and an excerpt (Missale abbreviatum) - of a missal which, like so many incunabula, contain no specification of printer, printing location or printing year, though for certain reasons it was suspected to have come from Gutenberg's printing office and was even regarded by several scholars as one of his earliest works, printed at the beginning of the 1440s. Since then three more copies have been discovered, one of them - the only really complete one - in 1925 in Zentralbibliothek, Zurich, by the Swedish national librarian Isak Collijn, and the third in Augsburg in 1961. All the time the dating of the book - the 1440s, the 1460s or c. 1480 - has been a subject of fierce controversy.

We have already remarked on the importance of sewing dots and states as criteria for the identification of one watermark with another. It was in fact Stevenson who properly established these concepts in filigranology and utilised them in his meticulous comparisons. Humorously and engagingly, he has described the time-consuming but exciting studies of the three Missale copies in Zurich, Munich and New York during the fifties, and the way in which the twin-pairs "Wideface" - "Squareface" and "Leveleye" -"Slanteye" (Stevenson's own name for these variants of Bull's Head, one of the two watermarks in the missal, the other being known as Cross-on-Mount or Drieberg) became a couple of good friends whose re-appearance always gave him the same pleasure, not least when they appeared in other places besides Missale Speciale! Stevenson was soon able to conclude, with almost absolute certainty, that this disputed work had been printed in 1473 or early 1474, but he postponed publication of his findings in order to make further additions to his documentation. Then, in 1960, exhaustive accounts were published in two specialised journals, announcing that two other paper researchers, quite unaware of each other's investigations and of the results obtained by Stevenson, had arrived at the same conclusion, namely that the paper in the missal dated from the first half of the 1470s!

In 1962 in Gutenberg-Jahrbuch, Stevenson presented the first report on his investigation, and in 1967 he published the document in full in his book "The problems of the Missale Speciale". During the intervening years he had obtained further confirmation of his findings by means of beta spectrometry.

With these entirely unanimous datings by three quite independent filigranologists, using three fairly distinct methods, the actual dating question would seem to be settled once and for all and filigranology can perhaps be said to have matriculated with flying colours. Contrary to the standpoint adopted by bibliographical research in 1951, the unexploited sources of information present in the watermarks of the paper were sufficient to solve the essential problem of Missale Speciale - its dating - without any further archival discoveries being necessary.

The dating of the Vinland Map is, first of all, not concerned with the map itself - which is drawn on parchment and, therefore, cannot contain a watermark - but rather with two documents, the Tartar Relation, found together with the map when the first find was made, and "Speculum historiale", written by Vincent de Beauvais and usually known as the Vincent Text, which is believed to have previously been bound together with the relation and the map (worm holes show as much) and, by an almost incredible coincidence, was discovered shortly after it. Secondly, the dating of the Bull's Head, which - as usual in the process of alternating with a "twin" - is practically the only watermark in both documents (one sheet presents a variant from another pair) is, filigranologically, a fairly simple business, as anybody with access to Briquet's Les Filigranes and Piccard's Ochsenkopfwasser-zeichen can see for themselves.

The exhaustive account of the find published in 1965 includes a letter from Stevenson (19.10.62) in which, referring to Briquet and to his own researches, he gives the following time and place of the watermark: c. 1440, probably from a certain papermill near Basle. This tallied extremely well with the results of the close investigations, already made at Yale, of all other criteria than watermarks. The book reproduces only two, identical, watermarks, one from the Tartar Relation and one from the Vincent Text. What we are shown, then, is only one of the twins, which has been regretted. The other, however, has been presented in specialist journals, e.g. in the above mentioned new edition of Les Filigranes

(1968), in which Stevenson reproduced betagrams of both twins (plus further evidence for them). Undeniably, in a sensational context like this, one would like to have many more confirmations and ample documentation as in his book on the Missale problem, and it is not unreasonable to suppose that, until his lamentable death, he was occupied on something of the kind.

The revolutionary consequences of this early dating combined with the many circumstances contradicting it guarantee that the Vinland Map will remain a subject of controversy for a long time to come. As has already been made clear, the dating of the map itself is entirely dependent on the confidence with which one can claim that it was originally united with the two documents. This seems to be amply confirmed, but it is not for the filigranologists to judge. He dates paper, nothing else.

Clearly, an adequate method of depiction is a prerequisite of watermark research, but there are occasions when depictions are not needed, when original watermarks can be compared with each other directly. This is not infrequently the case in Stockholm, where, in their old home on the island of Riddarholmen, the National Archives have a very large collection of unwritten original paper with watermarks, mostly half-sheets (Verso) which have been detached from documents in the course of earlier weeding-out operations. In terms of quantity this is definitely the world's biggest collection, but it is mostly confined to watermarks between 1650 and 1850. A good many of these, of course, are duplicates, but this in itself is all to the good, because duplicates are essential for the establishment of twins, states etc. The author of this article, who has been employed for many years in arranging and dating the collection, has had occasion to describe both the collection and his own work elsewhere.

The comparator is a small, handy device for ocular comparison of originals, of pictures or of originals and pictures. This is a mirror device in which the two items are made to coincide optically, which has the effect of heightening any differences between essential details.

The author feels prompted to round off this rhapsodical overview in the same words as he used in his first introductory essay on watermarks: "It has turned out to be both incomplete and prolix." This, however, is probably inevitable when one tries, in a limited space, to inform the common reader as clearly as possible about a completely unknown field of research.

Finally, we have been concerned here with the history of paper and filigranology solely as "handmaids of historical science", but of course they are of value in their own right as an independent field of knowledge - small, specialised but packed tight with interesting political, economic, social, cultural and - not least - fascinating aesthetic aspects which make them well worth studying for their own sake. Most writings on the history of paper, of course, concentrate on aspects of this kind. The fact of research in this field also

*Raphael's Madonna as a light-and-shade watermark.
Made by Pietro Miliani, Fabriano. From Hunter.*

generating results of interest to other branches of scholarship is all the more gratifying to its enthusiastic practitioners!

### *Art-handicraft and proof against forgery*

When the woven metal wire came into use, there ensued a veritable revolution in watermark technology. This not only made it possible for the metal wire profile to be more securely attached to the wire but also for the metal cloth itself to be used for watermarks. By embossing a picture or executing it in relief, one can produce a three-dimensional watermark. Starting in about the first years of the 19th century, this technique developed into a superb craft. Many paper-

235

236     *Combination of light and shade and thread watermark. From Silkeborg Paper Mill. Denmark.*

makers quickly mastered the technique, and one of the foremost of them all was Pietro Miliani of Fabriano in Italy. By combining the three-dimensional picture with the traditional wire and foil emblems, a paper was created which fills the layman with admiration and astonishment and the paper-maker with pride and happiness.

Fabriano, the oldest European papermill still in operation, has saved many tools from earlier times, and it is possible there to study the techniques which produced such exquisite watermarks several hundred years ago.

First the relief was carved in wood. The woven metal wire was then heated in a baker's oven and laid on the relief, and pushed into it with a folding stick or a peg, to make it conform exactly to the lines of the relief. Eventually the practice was adopted of cutting the relief in a thin piece of wood which was then glued onto a thicker piece. This technique is highly reminiscent of early typographic methods.

Another development consisted in modelling a relief in wax and making a bronze impression of it. This eventually led to the practice of making two bronze casts, a matrix and a patrix, and pressing the metal cloth between them. The result was a cloth with a relief pattern. This method, slightly modernised, is still being used today.

There was one problem involved in making portraits with this three-dimensional technique, also known as light and shade. Hand-made paper shrinks during drying by about 4.5 per cent of its length and 2.5 per cent of its width, and so the face of the finished work might not be recognisable. The mould-maker, therefore, has to allow for this and distort the wax model accordingly. This, as will be appreciated, demands immense technical skill, coupled with artistic sensibility. The light and shade technique also requires several more dips in the stock before the sheet can be couched. The Tumba vatmen used to dip the mould between twelve and twenty times before couching. Between each imersion, the mould was allowed to drain for an exactly timed period, so that the pulp would not slide off when the mould was plunged back into the vat.

Multi-coloured watermarks called for an even greater degree of skill on the part of the mould-maker and vatman. For every new colour, a new mould had to be made which was exactly similar to all the others. The Tumba papermakers also mastered this technique, which required several vats (one for each colour) and several vatmen working together. Each wet sheet had to be positioned exactly on top of the previous one, and even if jigs were used, this work required a great deal of precision.

# PAPER ART

**Liberating paper from its traditional limitations**

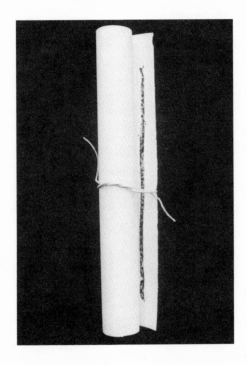

*Previous page: Rune Hagberg, Ikonoklasmos, 1977.*

The past 20 years have revealed a growth of interest among artists where paper is concerned, not as something to draw, paint or print on, but as a communicative medium in itself.

Making one's own paper has developed into a new art form, with experimental treatment of pulp to exact from it new expressive potentialities.

An immensely versatile material like pulp is also amenable to experiments in form together with other materials. Hand-made paper can also be adapted to artistic purposes in a completely different way from machine-made paper, partly because it does not have any grain direction. Machine-made paper is made almost entirely of pulp from different kinds of wood, either chemical or mechanical, whereas the hand-made paper-maker prefers rag pulp, if it is obtainable, or pulp made from cotton linters, abaca or other raw fibres.

This applies to occidental hand-made paper-makers. Japanese or other Asian paper-makers, with their distinctive technique, can make paper straight from the raw fibre, which is mostly bast from the paper-mulberry tree.

Many artists, by personally concerning them with the craft, have begun to take an interest in the production process and eventually begun making their own pulp - collecting and processing the fibres, diluting them to pulp, making their own sheets or using the pulp in multi-coloured compositions or sculptures.

In the infancy of modern art, for example, Picasso and Braque experimentally stuck pieces of wallpaper and fabric onto their paintings as part of the composition. Futurists, Dadaists and Surrealists also employed a great deal of collage in their works, but the real peak of achievement where collage is concerned came with Kurt Schwitter and his "Merz" paintings and objects. Rusty iron, weathered pieces of wood and bits of paper suddenly became important parts of the composition.

An elaboration of these ideas begins to appear in the early 1960s, with several artists concentrating on the surface structure of the work of art. Burri and Tapies with raw, torn and dyed pieces of fabric, Yves Klein with sponges and Fontana with slashes in the canvas. The first European and American experiments in paper art begin at about the same time. Several artists discover the potentialities of hand-made paper as an artistic medium, and eventually a completely new form of art comes into being. A form of art in which, it is true, paper is used, but as an independent medium instead of just a vehicle.

One artist who was quick to seize on the possibilities of paper for

242    *Rune Hagberg, to John Cage, 1976.*

experimentation was Rune Hagberg. The following excerpt from an article in Moderna Museets Tidskrift, the journal of the Swedish Museum of Modern Art, says a great deal about his attitude to the material:

"There is something remarkable about paper. There are dead papers - soulless! And there are inspired ones. Thin Chinese paper, made from tender bamboo, meant to be conveyed by carrier pigeons; Arabic paper treated with saffron; Persian paper smoothed by hand with a polished stone of onyx or agate; hard, firm, opaque European paper glazed between steel rollers; sun-bleached or chlorine-bleached paper; paper made of fibres from the cocoons of the silkworm, so fine that they seem on the verge of dissolving into nothingness; coarse paper made of ground pulp from massive logs pressed against rotating grindstones; paper in which fibres from the swaddling of Egyptian mummies have been re-used; paper treated with glue, kaolin, gypsum, chalk, rice starch, lime wash, fig juice; paper from Lou-lan, Tun huang, Khorasan, Danlatabad, the Escorial, Dordrecht, Troyes, Jean d'Heures, Lessebo, Klippan - they all have their individuality, their characteristics. Just as a wine is fashioned, not only by the manner of its preparation but also by the type of grape and the soil in which it is ripened, so paper depends on the species of fibre and the place where it grew.

"Paper to me is not just an anonymous background, an 'empty' ground material. Characters and paper belong together. There is a link here, it seems, with the watercolour art of the Romantic period, for example, when the paper, visible through the translucent brush strokes, had an important part to play. It had its intrinsic value, just as it has here in collage. Watercolour paper was fashioned with great care at the hand-made papermills, together for example with document, bank note and copper-printing paper.

"The paper machine soullessly churns out its long runs of exactly identical sheets. But because I, in my battle of this fibre field - a battle which proves, more and more, to be a struggle against the impossible - have come to fabricate them, handle them, maltreat them, every single sheet has acquired an individual value."

The individual value to which Rune Hagberg refers can be achieved in many ways. More and more artists are now starting with the pulp as their medium, building up their pictures from coloured stock. There are many ways of providing oneself with ready pulp. Natural fibres, such as flax, nettles, iris and lily leaves, do not give a very wide selection of colours; the range is from yellowish-green to medium-brown. If, on the other hand, one uses rag based on coloured cotton shirts, jeans and suchlike, then the existing colour can be used or else improved by means of batik or pigment paints.

One could not wish for a more docile material; the artist can make it himself in exact compliance with his own purposes.

The following cavalcade of paper art from Sweden, Denmark, Switzerland, Japan and the USA includes specimens of many techniques and procedures for achieving what is desired. From the very simplest method practised by Lottie Oleby (boiled straw and seaweed are beaten with mallets to the required consistency, after which the pulp is shaped into a three-dimensional picture) to more sophisticated Japanese and American methods in which the picture is built up in several layers of thin sheets.

The book is another thing in which artists have long been interested. The conception of a book hinges on many different factors - a combination of tactile and visual elements. In Flight Wings, Caroline Greenwald has created a book which is, in equal measure, a sculpture and an image of the book as an idea. But surely, a book which cannot be read is no book at all? Perhaps so, but I will nonetheless venture a definition: a book made for its own sake and not as a container for a lot of information. In other words, it does not contain a work - as, for example, in the case of a book of poetry - it is the work. Design and format reflect the content.

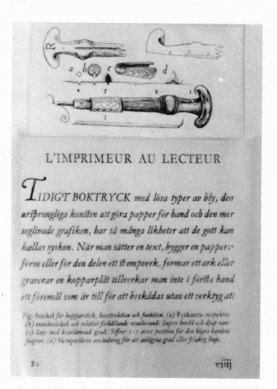

*An early print proof of Richard Årlin's book Libro de Horas, 1987.*

An artist's book of this kind is something quite different from an ordinary art book, which of course is just a retrospective exhibition of reproductions. This is not to disparage art books in general; they are beautiful and necessary as a documentation of artistic work.

There is also a procedure which is more similar to the old-fashioned production of books. Richard Årlin has worked a great deal with hand-made paper, both as a printer and, most recently, in a hand-made book, produced in 45 copies, where both the vision and the production are 100 per cent his own.

The idea of making a book originated some years ago, when Årlin came by an old, dented tin of pure cinnabar, suitable for printing initials and suchlike. This remarkably heavy can with its brilliant red contents gave rise to several years' work. Once Årlin had decided that the book was to deal with his experiences on a journey to Spain, the question arose of which paper to use. On his way to Spain, Årlin visited some old papermills in Auvergne, France, where a number of stamping mills had been preserved, and so he decided to build one of his own, so that he could make his own pulp and form his own sheets for the book-to-be.

Building the stamping mill took him about one year. It is entirely of wood - fine-textured pine - except for the trough, which is cast concrete. At the time of writing, Årlin has been running his stamping mill for 3 years without a hitch and has formed hundreds of sheets, among them 900 for the book. For the end papers he has composed a pulp consisting of linen and jute rags together with old jeans. The book, of course, is printed letter press, and printing, illustrations and binding Årlin's own work.

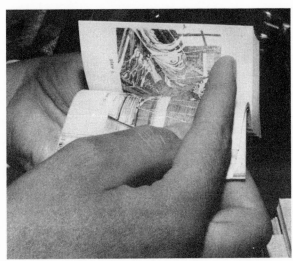

*One of Asao Shimura's hand-made books about paper.*

*Shifu weaver Kazuyo Kajiyama rolling kozo strips before twisting them into paper threads.*

Between these "extremes" - Greenwald's version of the book as an idea and object, and Årlin's classical artist's book, which has more in common with the books of William Blake a hundred years ago - there are thousands of other works, of equally personal design, in which paper, very often, plays an important part.

Asao Shimura in Japan is making a series of books entirely by hand, having paper and its history as their common denominator. The format is a classical Japanese one called mamehon (7.5 x 5 cm approx.) and Shimura is forming all the sheets, setting all text in lead type (Baskerville 6 points), printing the book on his small platen press and binding the little books himself.

Japan also has a craft called shifu. This can be briefly described as the art of weaving fabric from paper, but it is a highly complicated craft demanding great skill and considerable artistic talent.

Shifu has existed in Japan since the Edo period (1603-1867), but with the accelerating importation of cheap textiles at the turn of the century, demand fell off and the craft was virtually extinct until after the Second World War. Whereas previously it had been used for making garments and other objects of an everyday type, at the end of the 1940s a more experimental kind of production began, on a modest scale, but it is only in recent years that shifu has come in for the attention it deserves.

The input material for shifu is thin, tough kozo sheets, measuring about 60 x 90 cm. They are folded down the middle and then once more, so as to give an overlap of about 2 cm on each side. The sheets are then put on a special cutting board and cut in 2 or 3 mm narrow strips. The overlap is not cut through entirely, and so the sheet still hangs together.

The slit sheet is then rolled up in a damp towel, and left there for seven or eight hours. The strips then have to be rolled: this is considered the most difficult and delicate part of the whole process. The rolled-up, slit sheet is placed on a hard underlay, preferably stone, and then rolled to and fro.

After rolling, the strips have to be detached from the overlap, which is still holding them together, and then joined together in a single long thread. When this time-consuming, finicky task is completed, everything is immersed in warm water to keep the twisted ends from coming untwisted again.

The final stage in making shifu threads is to twist them once again on a spinning wheel rotated with the right hand, while the left hand keeps the thread tensioned and free from knots. One sheet yields about 160 m. thread, and it takes roughly a month to make enough thread for a kimono. One might not expect the paper fabric to be worth much after the washing or after a shower of rain, but the fact is that shifu is very strong and durable. It can even be machine-washed.

I hope this short article has made it clear what an immensely versatile material paper is. In fact the only limit is the user's imagination, so go right ahead and liberate paper from its limitations!

*Jane Balsgaard, Denmark. **Panama** 1990. 52x28x64 cm. Piassava, cochineal, stinging nettle, banana, madder root and birch leaves.*

*Yoichi Takada, Japan. **Tsubasa "Kei"** 1986. 230 x 215 cm. Kozo, bamboo, stone, iron, brass and lead.*

*Caroline Greenwald, USA. **Flight Wings** 1979. 42.5 x 65 cm. The cover of Mexican amate is lined with Japanese tea chest lining paper of silver. The series of pages are made of abaca pulp with tengujo and gampi papers incorporated with lines of monofilament and strands of white deer hairs.*

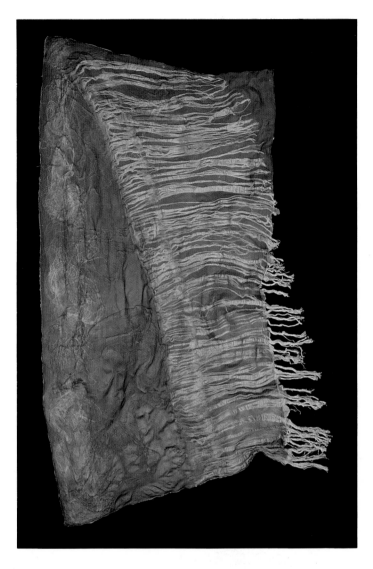

*Kerstin Svanberg, Sweden. **The Sun is blowing** 1989. 133x67x15 cm.
Linen pulp and threads. Photo by Jan Nordahl.*

*Anne Vilsbøll, Denmark.* **Screen sculpture** *1986. 135 x 110 cm. Pieces of plastic cast into hand-made abaca paper.*

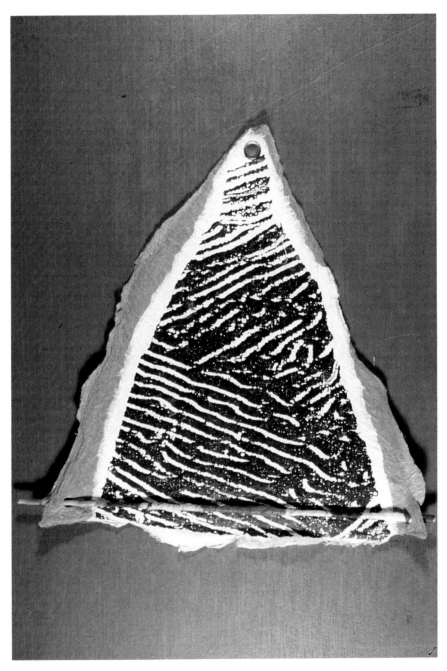

*Kazuyo Yamada, Japan.* **Waterland A** *1986. 27 x 22 cm. Kozo and mitsumata, bamboo stick, xerox and brass.*

*Isa Mitake, Japan. **Light Packaging** 1984. 180 x 60 x 45 cm. Kozo, lamp and metal frame, sesame oil.*

*Masaji Kasio, Japan. **The Big Ring** 1983. 182 x 180 x 22 cm. Iron wire, kozo, persimmon juice (tannin).*

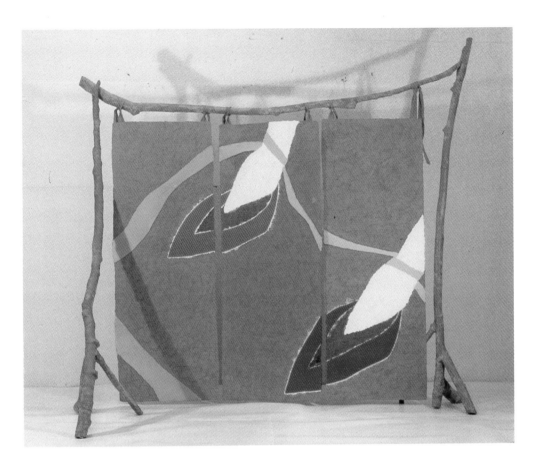

*Nicala Aiello, USA.* **The ties that bind** *1986. 8x6.5 feet. Handmade paper with colored pulp imagery, mixed media.*

*Fred Siegenthaler, Switzerland. **Waiting for ...** 1985. 43 x 38 cm. A thick layer of coloured cotton fibre was poured into the hand mould and covered with a second layer with dyed cotton fibres. The legs were embossed with a stencil made of board, when the cotton was still wet.*

*Bo Rudin, Sweden.* **Fontana's Japanese Cousin** *1982. 38 x 22 cm. Coloured cotton pulp mixed with newspaper pulp.*

*Lilian Bell, USA. **Shade and shard** 1985. 26x40x9 cm. Papers made with Japanese forming techniques, dried flat, then dyed with Indian ink and acrylic paint. After drying the sheets are dampened and formed over the three-dimensional objects and mounted into the frame.*

*Lottie Oleby, Sweden.* **Goldmund** *1985. 50 x 20 cm. Hand-beaten straw and newspaper pulp and wooden stick.*

*Akira Kurosaki, Japan. **Captured Time 84-3** 1984. 210 x 150 cm. Kozo and hemp paper, cotton thread, persimmon juice.*

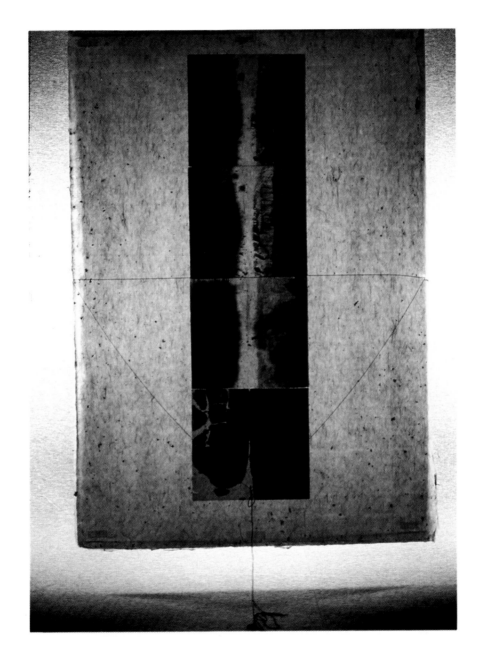

*Akira Kurosaki, Japan.* **Captured Time 84-1** *1984. 210 x 150 cm. Kozo and bamboo paper, sumi ink, cotton thread.*

*Shoichi Ida, Japan.* **Three Twigs No. 36** *1980. 90.5 x 68.5 x 11.5 cm. Lithography, drawing and a twig on moulded kozo paper.*

*Shoichi Ida, Japan.* **Water** *1981. 74.5 x 68.5 x 13. Watercolour on backside of moulded and piled kozo paper.*

*Pia Andersen, Denmark. **Spring 88** 1988. 40x40 cm. Collage with handmade paper of straw and abaca. Photo by Planet Foto.*

# BIBLIOGRAPHY
# &
# INDEX

*Althin, Torsten:* Papyrus 1895-1945. Göteborg 1945.
*Althin, Torsten:* Papyrus, Swedens largest fine paper and board mill. Stockholm 1953.
*Ambrosiani, Sune:* Dokument rörande de äldre pappersbruken i Sverige. Stockholm 1923.
*Ambrosiani, Sune:* Papperstillverkningen i Sverige intill 1800-talets mitt. Särtryck ur Molae Chartariae Suecanae. Stockholm 1923.
*Anstrin, Hans:* Papper och massa i svenskt folkhushåll. Stockholm 1949.
*Barrett, Timothy:* Japanese Papermaking. Tokyo 1983.
*Bell, Lilian A:* Papyrus, tapa, amate & rice paper. McMinnville 1983.
*Bell, Lilian A:* Plant Fibers for Papermaking. McMinnville 1981.
*Berg, Elisabeth:* Papper för hand. Stockholm 1986.
*Blum, André:* Les Origines du Papier. Paris 1932.
*Boethius, Bertil:* Grycksbo 1382-1940. Falun 1942.
*Bosaeus, Elis:* Munksjö Bruks minnen. Uppsala 1953.
*Bosaeus, Elis:* Silverdalen, historien om ett småländskt pappersbruk. Uppsala 1944.
*Briquet's Opuscula* (Dr Briquets samlade verk, med undantag av "Les Filigranes"). Hilversum 1954.
*Burg, Christer von der:* Japanskt handgjort papper. Stockholm 1976
*Carter, Jane Levis:* The Paper Makers. Pennsylvania 1982
*Castegren, Erik:* Riksbankens Pappersbruk, Tumba. Minnesskrift till dess tvåhundraårs-jubileum. Stockholm 1956.
*Christensen, Sigurd:* Papirbogen. Köpenhamn 1958.
*Classification and Definitions of Paper.* New york 1928.
*Clemensson Gustaf (redaktör):* En bok om papper. Uppsala 1944.
*Clemensson, Gustaf:* Klippans Pappersbruk 1573-1923. Lund 1923.
*Clemensson, Gustaf:* Lessebo 1658-1856. Stockholm 1954.
*Clemensson, Gustaf:* Lessebo 1856-1956. Stockholm 1964.
*Clemensson, Gustaf:* Papperets historia intill 1880. Uppsala 1953.
*Clemensson, Gustaf:* Svenska pappersbruksföreningen 1923-1948.
*Cox, Catherine:* Handmade paper in Tuckenhay, Devon. Ashburton 1977.
*Cross, C.F. & Bevan, E.J.:* A textbook of papermaking. London 1900.
*Ericsson, Harry & Lindström, Gösta:* Fröåsa Handpappersbruk. Virserum 1970.
*Eneroth, Otto:* Handbok i papperskännedom. Stockholm 1941.
*Eneroth, Otto:* Papper, dess framställning och användning. Stockholm 1934.
*Eurenius, C.A.:* Populär handledning i papperskännedom. Klippan 1923.
*Fiskaa Haakon:* Papiret og Papirhandelen i Norge i eldre tid. Oslo 1940.
*Garlock, Trisha:* Glossary of papermaking terms. San Francisco 1983.
*Gille, Torsten & Hansen Lars Ole:* Pappersbruksarbetareliv. Vimmerby 1981.
*Grafik-Nytt 1/2 1984.* Specialnummer om papper.
*Göth, J.A.:* Bergsrådet på Lessebo. Stockholm 1925.
*Hagen, Victor Wolfgang von:* The Aztec and Maya Papermakers. New York 1944.
*Handgjort Papper,* ett kompendium. Grafikskolan. Stockholm 1979.
*Hannover, H.I. & Smith, Sigurd:* Papirfabrikation. Köpenhamn 1934.
*Hassing, Oluf:* Papir. Köpenhamn 1947.
*Heller, Jules:* Papermaking. New York 1978.
*Helmfrid, Björn:* Holmenöden under fyra sekler. Norrköping 1954.
*Higham, Robert R.A.:* A Handbook of Papermaking. London 1963.
*Hockney, David:* Paper pools. London 1980.
*Hofmann, Carl:* Praktisches handbuch der Papier-Fabrikation 1-II. Berlin 1897.
*Hopkinson, Anthony:* Papermaking at home. Wellingborough 1978.
*Hughes, Sukey:* Washi, the world of japanese paper. Tokyo 1978.
*Hult, Nils:* Papper och pappersmassa. Stockholm 1969.
*Hunter, Dard:* Papermaking, the history and technique of an ancient craft. New York 1947.

Johansson, Albert & Eckerbom, Nils: Örebro Pappersbruk och dess historia. Stockholm 1951.

Karlsson, Kurt K.: Finlands handpappersbruk. Helsingfors 1981.

Larsson, Rutger: Pappersteknik. Hermods Korrespondensinstitut. U.å.

Lindgren, Torgny: Riksbankens sedelhistoria 1668-1968. Stockholm 1968.

Loeber, E.G.: Paper mould and mouldmaker. Amsterdam 1982.

Lundberg, Johan: Pappersmasseindustrien. Stockholm 1921.

Läsning för Sveriges allmoge, Första boken: Om skrifkonsten, boktryckerikonsten och papperet. Härnösand 1864.

Making Paper. Utställningskatalog. New York 1982.

Mason, John: Papermaking as an artistic Craft. Leicester 1963.

Michel, Albin: Machines pour la fabrication du papier. Paris 1922

Molae Chartariae Suecanae I-II. Svenska Pappersbruksföreningens Jubileumsskrift. Stockholm 1923.

Molin, Harry: Munkedal. En bygd och ett bruk i Bohuslän. Stockholm 1949.

Moon Boertzel, Barbara: Papermaking in Micronesia. Guam 1982.

Munksjö handbok. Stockholm u.å.

Narita, Kyofusa: A life of Ts'ai Lung and japanese papermaking. Tokyo 1980.

New American Paperworks. Utställningskatalog Wort Print Council 1982.

Nikander, Gabriel & Sourander Ingwald: Lumppappersbruken i Finland. Helsingfors 1955.

Nisser, Marie & Sjunnesson, Helene: Massafabriker och pappersbruk i Värmland och Dalsland. Stockholm 1973.

Nyström, Karl: Den norrländska trämasse- och cellulosaindustrin. Uppsala 1924.

Paper and Printing: The new technology of the 1830's. Oxford 1982.

Paper - art and technology. World print council 1979 Washington.

Paper as Image. Utställningskatalog. London 1983.

Paper Makers Association: Paper Making. Surrey 1949.

Paper Trails Utställningskatalog. Liverpool 1985.

Papier als Künstlerisches Medium. Utställningskatalog. Zweibrücken 1980.

Papier erzählt, Die Geschichte einer Papiermühle am Teutoburger Wald. Bielefeld 1949.

Papir Bogvennen april 1976. Köpenhamn 1976.

Papper. Utställningskatalog. Röhsska Konstslöjdsmuseet. Göteborg 1982.

Papper. Utställningskatalog. Grafiska Sällskapet. Stockholm 1983.

Pappersmasseförbundet 1907-1957. Stockholm 1957.

Pappersordlista TNC 66. Stockholm 1977.

Papyrus. Skrift med anledning av bolagets första tjugofemårsperiod. Göteborg 1921.

Richardson, Maureen: Plant Papers. Essex 1980.

Rischel, Anna-Grethe: Det klassiske papirmageri i Nepal. Saertryk af Nationalmuseets Arbejdsmark 1985.

Siegenthaler Fred: Just paper. Düren 1986.

Sköld, Birgit & Turner, Silvie: Handmade Paper today. London 1983.

Smith, Sigurd: The action of the beater. London 1923.

SPCI: Cellulosa och papper. Stockholm 1948.

SPCI: Ordlista för massa, papper, fiberskivor. Stockholm 1958.

Sporhan-Krempel, Lore: Chronik der Papier-Macherei im Raum Osnabrück. Osnabrück 1958.

Stephenson, J. Newell: Pulp and paper manufacture vols. I-IV. New York 1950.

Stevenson, Allan H.: Briquet and the future of paper studies. Hilversum 1955.

Studley, Vance: The art and craft of Handmade Paper. New York 1977.

Svenska Pappers- och cellulosaingenjörsföreningen - 25 år. Stockholm 1933.

Tjerneld, Staffan: Munksjö - en bildsvit. Stockholm 1962.

Toale, Bernard: The Art of papermaking. Worcester, Mass. 1983.

*Trier, Jesper:* Ancient paper of Nepal. Århus 1972.

*Tsien Tsuen-hsuin:* Paper and printing (vol 5:1 of Needham, Science and Civilisation in China). Cambridge 1985.

*Tsichold, Jan:* Ts'ai lun, Papirets opfinder. Köpenhamn 1957.

*Waldén, Bertil:* Frövifors Bruk och dess föregångare Frövi Bruk. Stockholm 1951.

*Valenta, Eduard:* Das Papier. Halle 1922.

*Westberg, Hans:* Östanå Pappersbruk i Hälsingland. Sundsvall 1965.

*Vestergren, Johannes:* Papperstillverkning. Stockholm 1924.

*Vilsbøll, Anne:* Papirmageri. Köpenhamn 1985.

*Works Paper.* Utställningskatalog. Ohio 1982.

*Öman, Erik:* Cellulosaindustrien. Stockholm 1944.

277

Bo Rudin MAKING PAPER
© Bo Rudin 1990
Translated from Swedish by Roger G. Tanner
Typeset by Lillan Östlund
Printed by Bildtryck, Stockholm
Jacket Illustration *Couching* by Bo Rudin
Paper: *Mischa* from Forenede Papir, Copenhagen
*Macoprint* from Pappersgruppen, Stockholm (colour sheet)
*Ingres* jacket and *Tre Kronor* (end paper) from Tumba Bruk.
Edited and designed by RUDINS Publishers, Box 5058, S-162 05 VÄLLINGBY, Sweden
ISBN 91-970-8882-X